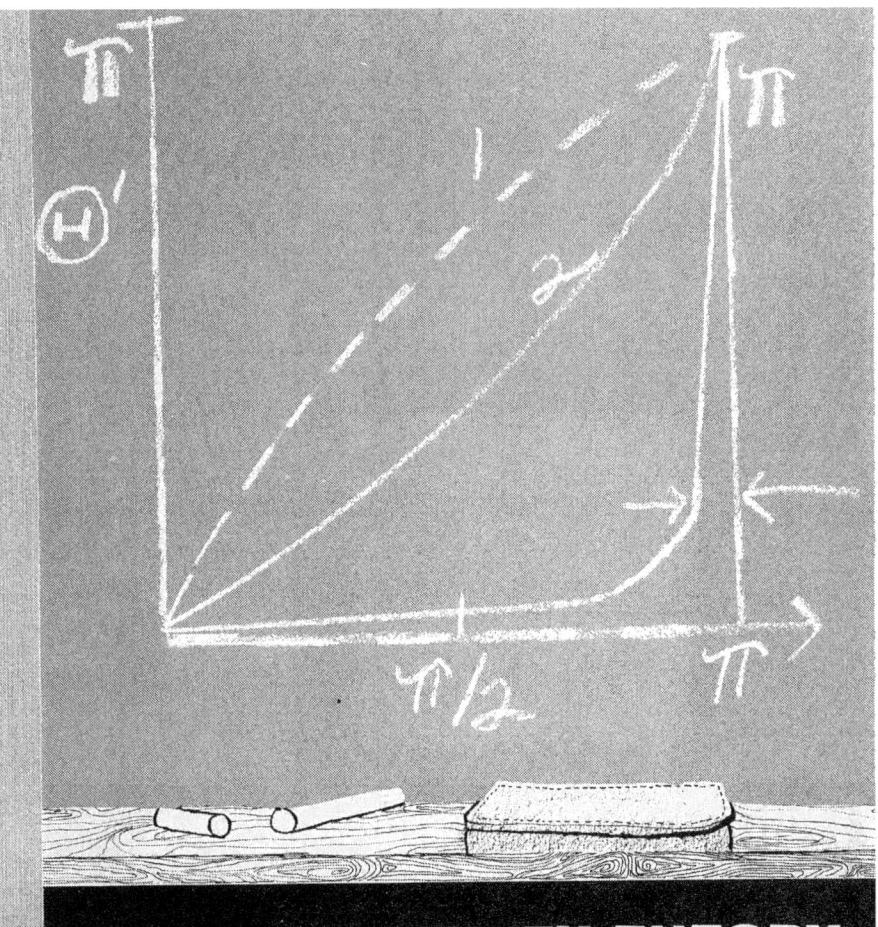

SPECIAL RELATIVITY THEORY

ELECTED REPRINTS

ished for the AMERICAN ASSOCIATION of PHYSICS TEACHERS

he AMERICAN INSTITUTE of PHYSICS · 335 East 45th Street, New York 17, N.Y.

Resource Letter SRT-1 on Special Relativity Theory †

(January 1962)

Prepared by Gerald Holton, Department of Physics, Harvard University, at the request of the Commission on College Physics.

This is one in a series of Resource Letters on different topics, intended to guide college physicists to some of the literature and other teaching aids (e.g., films, educational apparatus) that may help them to improve course contents in specified fields of physics. No Resource Letter is meant to be exhaustive and complete; in time, there may be more than one on each of the main subjects of interest. Your suggestions and comments will be welcomed.

Terminology: The letter E after the consecutive item number (e.g., 2.E) means the items should be mainly useful for *elementary* (freshman liberal arts through sophomore physics) courses; the suffix I (e.g., 7.I) indicates *intermediate* (junior, senior) courses; and the suffix A indicates *advanced* (senior, graduate) courses. An asterisk (*) signals items particularly recommended for introductory study.

Additional copies: Available from American Institute of Physics, 335 East 45 Street, New York 17, New York. When ordering, request Resource Letter SRT-1, and enclose a stamped return envelope.

I. INTRODUCTION

Over the next few years, a good deal more SRT (special relativity theory) will find its way into introductory and intermediate college physics courses. With this aim in view, plans for writing and for equipment development are at last under way in several colleges. Instructors everywhere should seriously think now of participating in this effort, or at least of getting themselves and their classes ready to use more teaching material on SRT. For this purpose, the present Resource Letter, the first one on this subject, takes a selective look at certain helpful resources that are now available on some of the main topics in SRT. It is intended to be used by instructors, but may also be useful to students doing a course essay on a specific subtopic.

The instructor interested in introducing more SRT in his course faces at least five problems: (1) The amount of publication is immense, but its usefulness for the classroom is usually low. (2) Very little equipment or other aids exist to help the man behind the empty lecture bench make his points. (3) Enough time must be made for SRT in the course if it is to have any meaning to the student. (4) A clear line of argument for getting through SRT (and possibly to introductory GRT) must be selected from among several possibilities. (5) Enough integration with other topics (e.g., QM, nuclear physics) must be achieved to do justice to the full power of SRT.

This Resource Letter should help to deal with the first of these problems, that of the huge bulk of publications. No doubt the best way to begin using the Resource Letter is to find among the next eight sections one on which to concentrate.

† Preprinted from the American Journal of Physics.

II. BOOKS: SOURCES AND TEXTS

Seven books are so good, so basic, and so widely used that nothing new need be said about them. They are, so to speak, double-starred, and must be nearby and accessible if one teaches SRT: A. S. EDDINGTON, The Mathematical Theory of Relativity; A. EINSTEIN, Relativity, the Special and General Theory; A. EINSTEIN, The Meaning of Relativity; M. VON LAUE, Die Relativitätstheorie; C. MØLLER, The Theory of Relativity; R. C. TOLMAN, Relativity, Thermodynamics, and Cosmology; H. WEYL, Space, Time, Matter.

*1.I The Principle of Relativity. A Collection of Original Memoirs on the Special and General Theory of Relativity. A. EINSTEIN, H. A. LORENTZ, H. MINKOWSKI, AND H. WEYL. Notes by A. Sommerfeld. (Dover Publications, New York, 1958), 216 pp. (Reprint of 1923 translation.) Paperback. The translation of Einstein's basic paper of 1905 in this edition is much better than another translation prepared in 1920 by Saha and Bose. Among detailed analyses and critiques of content of this paper, see R. Dugas (item 29) and L. Silberstein (item 39).

*2.E La Cinématique relativiste. H. ARZELIÈS. (Gauthier-Villars, Paris, 1955), 228 pp. Among the most careful and readable works, with extensive bibliography for each chapter. This book and item 3 should be a major new resource for instructors.

*3.I La Dynamique relativiste et ses 'applications. H. ARZELIÈS. (Gauthier-Villars, Paris.) Vol. I, 1957, 304 pp.; Vol. II, 1959, 451 pp. See item 2.

*4.E Einstein's Theory of Relativity. MAX BORN. (E. P. Dutton and Company, Inc., New York, 1924), 293 pp. Without doubt the best elementary account. Thoroughly works out everything, from how to plot graphs through Maxwell, Minkowski, to GRT, with

28675

very few rabbits being pulled out of the hat. Leaving it so long out of print is a black mark against the world of publishing, but a paperback edition is now said to be planned.

5.I **La Théorie de la relativité restreinte.** O. COSTA DE BEAUREGARD. (Masson et Cie, Paris, 1949), 174 pp. Bibliography develops and uses tensors from the beginning.

*6.I **Introduction to the Theory of Relativity.** P. G. BERGMANN. (Prentice-Hall, Englewood Cliffs, New Jersey, 1942), 287 pp. Long a standard text. Introduces tensor calculus but uses it sparingly in Part I (SRT). With problems. Bergmann has been announced to have written the section on SRT in the new *Handbuch der Physik, IV*, which may turn out to be useful in place of a new edition of this book.

7.I **Relativity Physics.** W. H. McCREA. (Methuen, London, and John Wiley & Sons, Inc., New York, 1954), 4th ed., 87 pp. Useful condensation of the essentials, with footnotes (though mostly only to work before 1936). Herbert Dingle's elementary companion monograph, *The Special Theory of Relativity* (Methuen, London, and John Wiley & Sons, Inc., New York, 1940), 91 pp., is said by Dingle to "serve as an introduction" to McCrea's.

*8.I **Theory of Relativity.** W. PAULI. (Pergamon Press, New York, London, 1958), revised ed., 241 pp. Originally "a complete review of the whole literature on relativity theory" up to 1921, it was brought up to date by 26 supplementary pages written in 1956 (though these are mostly on GRT). This book still functions as a major *Handbuch* for its field.

9.I **Mathematics of Relativity.** G. Y. RAINICH. (John Wiley & Sons, Inc., New York, 1950), 173 pp. "Old Physics," "New Geometry," "Special Relativity," "Curved Space," and "General Relativity." Aims to bring "complete clarity" to presentations of SRT "by stressing mathematical aspect of the subject," introducing the more sophisticated tools step by step. Exercises.

*10.I **Special Relativity.** W. RINDLER. (Oliver and Boyd, Edinburgh and London; Interscience Publishers, Inc., New York, 1960), 179 pp. A concise, well thought-out book; assumes only elementary calculus and vector theory, develops tensors as needed. Numerous exercises. Proceeds via Lorentz transformation, kinematics, optics, tensors, mechanics of mass points, electrodynamics.

*11.I **Special Relativity for Physicists.** G. STEPHENSON AND C. W. KILMISTER. (Longmans Green and Company, London and New York, 1958), 108 pp. Short but thorough. Written particularly for the experimental physicist, with a large number of specific applications. No tensor calculus. Short bibliographies.

12.I **Les Principes de la théorie électromagnétique et de la relativité.** MARIE-ANTOINETTE TONNELAT. (Masson et Cie, Paris, 1959), 394 pp. Clearly written, orderly, with full derivations, descriptions of experiments, and references to original papers. Part

I, Electromagnetic Theory. Part II, SRT. Part III, GRT.

13.A **Spezielle Relativitätstheorie.** ACHILLES PAPAPETROU. (VEB Deutscher Verlag der Wissenschaften, Berlin, 1955), 170 pp. Two chapters (LT and Minkowski, dynamics of mass point) without tensors; three (electrodynamics, mechanics of continua, conservation laws) with tensors.

14.A **Relativity: The Special Theory.** J. L. SYNGE. (North-Holland Publishing Company, Amsterdam, 1956), 450 pp. ". . . the essentials of relativity from the Minkowskian point of view. . . . My ambition has been to make space-time a real workshop for physicists, and not a museum visited occasionally with a feeling of awe." Hence original and stimulating. Bibliography. See also his similar but short treatment, "Relativistic dynamics," in Vol. III, Part I (pp. 198–225) of Handbuch der Physik, S. Flügge, editor (Springer-Verlag, Berlin, 1960). Extensive bibliography of books on mechanics which treat SRT.

III. BOOKS: POPULAR

15.E **Space and Time.** EMILE BOREL. (Dover Publications, New York, 1960.) Original publication, 1922. 230 pp. Paperback. Still a useful, elementary book.

16.E **La Relativité.** PAUL COUDERC. (Presses Universitaires de France, Paris, 1958), 136 pp. No. 37 in series *"Que sais-je?"*, Paperback. A translation of this fine book has been arranged.

17.E **Readable Relativity.** CLEMENT V. DURELL. (G. Bell and Sons, London, 1926, and Harper and Brothers, New York, 1960, TB 530), 146 pp. Paperback. A passion for Lewis Carroll is offset by straightforward approach and exercises.

18.E **Space, Time, and Gravitation.** A. S. EDDINGTON. (Harper and Brothers, New York, 1959, TB 510.) Reprint of 1920 edition. 213 pp. Paperback. Mostly on GRT.

*19 E **The Evolution of Physics.** ALBERT EINSTEIN AND LEOPOLD INFELD. (Simon and Schuster, Inc., New York, 1938), 319 pp. A classic, now also in paperback.

*20.E **Ideas and Opinions.** ALBERT EINSTEIN. (Crown Publishers, Inc., New York, 1954), 377 pp. Paperback. The best general collection of Einstein's essays. Part 5 is "Contributions to Science."

21.E **What is Relativity?** L. D. LANDAU AND G. B. RUMER. (Oliver and Boyd, Edinburgh and London, 1960). Translation of 1959 book. 64 pp. Paperback. A "lighthearted presentation," very qualitative.

22.E **The Einstein Theory of Relativity.** LILLIAN R. LIEBER. (Reinhart and Company, New York, 1945), 324 pp. Surprisingly sophisticated.

23.E **Die Idee der Relativitätstheorie.** HANS THIRRING. (Springer-Verlag, Vienna, 1948), 3rd ed. 168 pp. Paperback. Qualitative only.

Note: See also M. Born, item 4, and Einstein, **Relativity, the Special and General Theory.**

IV. BOOK CHAPTERS

A few introductory general texts and most intermediate-level general texts have a section on SRT. Necessarily, most of these sections are rather similar to one another. The following list contains mostly recent, largely intermediate-level books that have been selected as being among those that it would be profitable to look through, particularly for illustrations and problems.

24.E **Principles of Modern Physics.** A. P. FRENCH. (John Wiley & Sons, Inc., New York, 1958), Chapter 6, pp. 137–173, "Relativity," emphasizes experiments and graphical presentations.

25.E **The Basic Concepts of Physics.** C. W. SHERWIN. (Holt, Rinehart, and Winston, Inc., New York, 1961), Chapter 4, pp. 82–151, "Relativity," is a refreshing discussion using ingenious little drawings.

26.E **Physics of the Atom.** M. RUSSELL WEHR AND JAMES A. RICHARDS, JR. (Addison-Wesley Publishing Company, Reading, Massachusetts, 1960), Chapter 5, pp. 111–145. With simple, worked-out examples and 27 problems.

27.E **Elementary Modern Physics.** RICHARD T. WEIDNER AND ROBERT L. SELLS. (Allyn and Bacon, Inc., Boston, 1960), Chapter 2, pp. 43–86, "The Theory of Special Relativity." Patiently carried out; 40 problems.

28.I **Theorie der Elektrizität. Band II.** R. BECKER. (B. G. Teubner, Leipzig, Germany, 1933), 6th ed., 400 pp. Useful discussion of optical experiments, pp. 255–265, and "Die Mathematischen Hilfsmittel der Relativitätstheorie," pp. 282–296.

29.I **A History of Mechanics.** RENÉ DUGAS. (Editions du Griffon, Neuchatel, Switzerland, and Central Book Company, New York, 1955), Part 5, Chapter 1, pp. 463–501, "Special Relativity." Despite a poor translation, this is an interesting analysis of the early work on SRT by Poincaré, Einstein, Minkowski, and Painlevé.

30.I **Classical Mechanics.** HERBERT GOLDSTEIN. (Addison-Wesley Publishing Company, Inc., Reading, Massachusetts, 1950), 399 pp. Chapter 6, pp. 185–214, "Special Relativity in Classical Mechanics," a very clear treatment. 14 exercises.

31.I **Principles of Modern Physics.** ROBERT B. LEIGHTON. (McGraw-Hill Book Company, Inc., New York, 1959), Chapter 1, pp. 1–56, "The Theory of Relativity." Brief but thorough. Simple tensor treatment. 86 exercises.

32.I **Introduction to Modern Physics.** F. K. RICHTMYER, E. H. KENNARD, AND T. LAURITSEN. (McGraw-Hill Book Company, Inc., New York, 1955), 5th ed., 666 pp. Chapter 2, pp. 49–76, "The Theory of Relativity," in briefest compass.

33.A **The Theory of Space, Time, and Gravitation.** V. FOCK. (Pergamon Press, London and New York, 1959), 411 pp. "The main purpose of this book was to develop the theory of gravitation from a new point of view," but even Chapter 1, on the special theory of relativity, is stimulatingly different from usual treatments.

34.A **A History of the Theories of Aether and Electricity. Vol. II. The Modern Theories, 1900–1926.** SIR EDMUND WHITTAKER. (Philosophical Library, Inc., New York, 1954). (Also Harper Torchbook TB 532). Chapter II, "The Relativity Theory of Poincaré and Lorentz," is a useful, brief account of the development of ideas up to Schrödinger's formulation of Minkowski's energy tensor. Copious citations in footnotes serve well for finding original papers. The bias indicated in the title of the chapter is analyzed in G. Holton, "On the Origins of the Special Theory of Relativity." Am. J. Phys. 28, 627–636 (1960). See also Vol. I, subtitled *The Classical Theories* (Harper Torchbook TB 531), Chapter XIII, "Classical Theory in the Age of Lorentz."

35.I "Relativity and electrodynamics." W. F. G. SWANN. Revs. Modern Phys. 2, 243–304 (1930). Listed here because it is still better than many book chapters. A thoughtful and well-written review-introduction to SRT stands up surprisingly well after 32 years.

Note: Useful chapters dealing with special topics exist also in other general texts such as Panofsky-Phillips (item 44) ; Shankland (item 36) ; R. B. Lindsay and H. Margenau, **Foundations of Physics;** W. Band, **Introduction to Mathematical Physics;** and A. Sommerfeld, **Lectures on Theoretical Physics.**

V. SELECTED EXPERIMENTAL WORK

The fundamental significance of SRT does not depend on any one of the many experiments to "prove" SRT. Still, the larger part of contemporary physics, from spectroscopy to accelerator design, is concerned with experiments and theoretical arguments that do rest at some point on SRT. For introductory purposes, one may well discuss a few experiments, one or two each from, say, optics and particle dynamics.

It will be helpful to keep in mind a classification of the various experiments having a bearing on SRT, and to be aware of their chronological sequence. One such scheme (based in part on a suggestion by D. L. Livesey) follows here. In searching for more complete listings of original papers, see the bibliographical citation in Whittaker (item 34), Pauli (item 8), Arzeliès (items 2, 3), von Laue's Relativitätstheorie, Thirring (item 38). A magnificent listing of all work prior to 1924, ordered by author's name as well as chronologically, is Bibliographie de la relativité, Maurice Lecat (Lamertin, Brussels, 1924).

Aether-drag in dense media: (a) Convection effects (e.g., Arago, 1818; Fizeau, 1859; MM, 1886; Zeeman, 1914). (b) Null experiments (Hoek, 1868; Airy, 1871). (c) Drag by large bodies (Lodge, 1893; Michelson, 1897).

NOTES AND DISCUSSION

Aether-drift experiments (air or vacuum): (a) First-order effects (Cedarholm *et al.*, 1958). (b) Second-order effects (Michelson, 1881; MM, 1887; Kennedy-Thorndike, 1932; Essen, 1955). (c) Indirect aether-drift effects (Double refraction: Rayleigh, 1902; Brace, 1904. Torque on condensor: Trouton-Nobel, 1903; Chase, 1926–27; Tomaschek, 1925–27).

Effects of moving source or mirror: (de Sitter, 1913; Michelson, 1913; Majorana, 1917–19; Tomaschek, 1924).

Rotating frame experiments: (Sagnac, 1913–14; Michelson, 1925; Ditchburn-Heavens, 1952).

Time dilation: (a) Transverse Doppler effect (Ives-Stilwell, 1938, 1941; Otting, 1939) (b) Lifetime of mesons (Rossi-Hall, 1941; Durbin-Loar-Havens, 1952).

Dynamics of high-speed particles: (a) Deflection of electrons (Bucherer, 1908–09; Neumann, 1914; Guye-Lavanchy, 1916; Rogers *et al.*, 1940). (b) Fine structure of H lines (Sommerfeld, 1916; Glitcher, 1917; Williams, 1938). (c) Scattering (Compton effect: Compton, 1923. Electron-electron scattering: Champion, 1932; Joliot, 1935; Leprince-Ringuet, 1936. Proton-proton scattering: Chamberlain-Segré, 1952).

Mass-energy equivalence in nuclear physics: (a) Mass defect (Cockcroft-Walton, 1932; Bainbridge, 1933). (b) Pair production (Anderson, 1933; Blackett and Ochialini, 1933). (c) Annihilation radiation (Klemperer, 1934; DuMond *et al.*, 1949). (d) Decay schemes of elementary particles (Marshak, 1955). (e) Other (e.g., Linac limiting velocity, Mössbauer effect).

A. General Surveys

*36.I Atomic and Nuclear Physics. ROBERT S. SHANKLAND. (The Macmillan Company, New York, 1960), 2nd ed., 665 pp. A clear, well-illustrated and experimentally oriented book, selected here as representative of the best ones for the purposes of this Resource Letter. References to original papers. Among experiments discussed are some on fine structure of spectral lines, electron mass variation with velocity, beta-ray spectra, synchrocyclotron design, meson production, binding energies, fission and fusion reactions, nucleon scattering, and lifetime of mesons.

37.I "Quelques vérifications expérimentales récentes de la théorie de la relativité restreinte." ROBERT LENNUIER. Rev. Sci. (Paris) 85, 740–748 (1947). A good, brief survey, with main results. The partial list, showing size of field to chose from, includes fine structure of hydrogen lines, Ives-Stilwell experiment, lifetime of mesons measured by Rossi *et al.*, effects of particle speed on inertia and on scattering angle (details of Leprince-Ringuet's work), and Q values.

38.I Handbuch der Physik. Band XII: Theorien der Elektrizität. (Verlag Julius Springer, Berlin, 1927), 564 pp. Chapter 3, pp. 245–348. "Elektrodynamik bewegter Körper and Spezielle Relativitätstheorie,"

by H. THIRRING. A useful summary, with accent on experimental results. Full bibliographical notes.

39.I The Theory of Relativity. L. SILBERSTEIN. (Macmillan and Company, Ltd., London, 1924), 563 pp. Still useful, both for detailed derivations and accounts of experiments.

40.I Kritik und Fortbildung der Relativitätstheorie. KARL SAPPER, editor. (Akademische Druck-u. Verlagsanstalt, Graz, Austria, 1958), 283 pp. Essays by Benedicks, Giese, Golling, Grünbaum, Mohorovičić, Moon, Sapper, Spencer, Tonini, Wenzl, and Zinsen. "Agreement on the need for rejecting dogmatism unites the otherwise widely differing points of view of the authors." Chosen here to stand for the many books critical of Einstein's RT; contains attacks on the usual interpretation of experimental results and on "the dangerous influence of positivism in physics."

Note: In addition, surveys of the main experiments are given in some of the more general books, particularly Arzeliès (items 2, 3), Tonnelat (item 12), Panofsky-Phillips (item 44), Stephenson and Kilmister (item 11), and Rindler (item 10).

B. Optical Experiments: Reviews

41.E Grundversuche der Physik in historischer Darstellung. CARL RAMSAUER. (Springer-Verlag, Berlin, 1953), 189 pp. This very useful book contains (pp. 63–70) quantitative descriptions of the main experiments by Roemer, Fizeau, and Foucault. Bibliography.

42.E Historic Researches. Chapters in the History of Physical and Chemical Discovery. T. W. CHALMERS. (Charles Scribner's Sons, New York, 1952), Chapter 4, pp. 64–83, "The Ether Drift Experiments," describes several experiments qualitatively. Slightly marred by somewhat naive philosophizing.

43.E A Source Book in Physics. W. F. MAGIE. (McGraw-Hill Book Company, Inc., New York, 1935). Brief excerpts from publications on velocity of light measurements by Michelson and Morley (1887; pp. 369–377), as well as by Roemer, Bradley, Fizeau, and Foucault (pp. 335–345).

*44.I Classical Electricity and Magnetism. WOLFGANG K. H. PANOFSKY AND MELBA PHILLIPS. (Addison-Wesley Publishing Company, Inc., Reading, Massachusetts, 1955), 400 pp. Chapter 14, pp. 230–242, "The Experimental Basis for the Theory of Special Relativity," is as pithy and useful a treatment as exists. Includes the table comparing trials of the MM experiment, from R. S. Shankland *et al.* (item 53) and a discussion of Kennedy-Thorndike experiment. Bibliographies to original experimental papers. (See also the chapters from 15 on—for other valuable teaching resources on SRT.)

45.A "Postulate versus Observation in the Special Theory of Relativity." H. P. ROBERTSON. Revs. Modern Phys. 21, 378–382 (1949). The task set here is that of "replacing, so far as possible, Einstein's relativity postulate by facts drawn from experience."

This is done by discussion of MM, Kennedy-Thorn-dike, and Ives-Stilwell experiments.

C. Optical Experiments: Largely on MM

*46.E **Light Waves and Their Uses.** A. A. MICHELSON. (University of Chicago Press, Chicago, 1903), 166 pp. Describes his experiments to 1899. Note: Bibliographies for the various M and MM experiments and discussions concerning them are given, e.g., in Whittaker (item 34); M. von Laue, **Relativitätstheorie**; Lecat, **Bibliographie de la relativité**; Jaffe (item 48).

47.E **Studies in Optics.** A. A. MICHELSON. (University of Chicago Press, Chicago, 1927), 176 pp. Contains quantitative descriptions of experiments.

48.E **Michelson and the Speed of Light.** BERNARD JAFFE. (Anchor Books, Doubleday and Company, New York, 1960). Describes the experiments qualitatively. Bibliographies.

49.E **The Principle of Relativity.** E. CUNNINGHAM. (Cambridge University Press, New York, 1914), Chapter 2, pp. 15–20. Consideration of reflection of light at a moving mirror in MM experiment. Same material in Cunningham, **Relativity and the Electron Theory** (Longmans Green and Company, London, 1915), pp. 16–20. More sophisticated treatments in Silberstein (item 39) and in references given there.

50.I **Die Physik 1914–1926.** O. D. CHWOLSON. (F. Vieweg und Sohn, Braunschweig, Germany, 1927), Chapter 16, "Der neue Versuch von Michelson (1925)." Detailed description and analysis.

51.I **"Conference on the Michelson-Morley Experiment."** Astrophys. J. 68, 341–402 (1928). A general examination of methods employed and results obtained. Contributions by Michelson, Lorentz, Miller, Kennedy, Hedrick, Epstein, and Bateman. Also a discussion.

*52.E **"Ether-drift Experiments and the Determination of the Absolute Motion of the Earth."** D. C. MILLER. Revs. Modern Phys. 5, 203–242 (1933). A good fullscale review, with photographs of equipment, bibliography, etc.

53.A **"New Analysis of the Interferometer Observations of Dayton C. Miller."** R. S. SHANKLAND, S. W. McCUSKEY, F. C. LEONE, AND G. KUERTI. Revs. Modern Phys. 27, 167–178 (1955). The famous case study on the effects of small experimental uncertainties; still reads like a fine detective story.

54.I **"A New Aether-Drift Experiment."** L. ESSEN. Nature 175, 793–794 (1955). A fairly rough MM experiment using microwaves (9200 Mcs).

Note: See also items: 1 (Lorentz on M experiment, pp. 1–7), 25, 34 (Vol. I, 390–2, 404–5), 49, 39, 43, 44, 45, 28.

D. Variation of Electron Mass with Velocity

55.E **Lectures on Theoretical Physics.** H. A. LORENTZ. (Macmillan, London, 1931), Vol. III, Chapter 7, pp. 269–288. "Experimental Investigations on the Mass of the Electron." On Kaufmann, Bestelmeyer, Buch-erer, Neumann, Guye, and Lavanchy, with data and descriptions. For more recent, briefer survey of this ground, see item 36.

56.E **Elements of Modern Physics.** PAUL L. COPELAND AND WILLIAM E. BENNETT. (Oxford University Press, New York, 1961), pp. 60–72, "Experimental Verifications of the Equivalence of Mass and Energy," contains detailed description of Perry and Chaffee experiment on e/m for electrons by time-of-flight measurement.

57.I **"Messungen der Massenveränderlichkeit des Elektrons an schnellen Kathodenstrahlen."** M. NACKEN. Ann. Physik 23, 313–329 (1935). Electron beam accelerated to nearly 0.7 c.

58.E **"A Determination of the Masses and Velocities of Three Ra-B Beta-Particles."** M. M. ROGERS, A. W. McREYNOLDS, AND F. T. ROGERS, JR. Phys. Rev. 57, 379–383 (1940). Check of relativistic formula to speeds up to 0.75 c, to distinguish between Abraham and Lorentz models of electron.

59.I **"Review of the Experimental Evidence for the Law of Variation of the Electron Mass with Velocity."** P. S. FARAGÓ AND L. JÁNOSSY. Nuovo cimento (Series 10) 5, 1411–1436 (1957). An extensive and skeptical review of main experiments to that date. See also letter by Champion, Nuovo cimento 7, 122 (1958); and counterargument, Raboy and Trail, Nuovo cimento 10, 797–803 (1958).

Useful graphs giving values of m/m_0 and β versus accelerating potential are given in S. Dushman, "Mass-energy relation," Gen. Elec. Rev. 47, 6–13 (1944).

Note: See also theoretical papers in Section VI below.

E. Some Other Representative Experiments

*60.I **"New Experimental Tests of Special Relativity."** J. P. CEDARHOLM, G. F. BLAND, B. L. HAVENS, AND C. H. TOWNES. Phys. Rev. Letters 1, 342–343 (1958). Comparison of frequencies of two masers having their beams of NH_3 molecules travel in opposite directions; result is that "ether drift" would have to be below 10^{-4} of earth's orbital velocity. [The effect is of first order in velocity of laboratory with respect to the ether. Theory in C. Møller, Suppl. Nuovo cimento 6, 381–398 (1957).]

61.I **"Nuclear Dynamics, Experimental."** M. STANLEY LIVINGSTON AND HANS A. BETHE. Revs. Modern Phys. 9, 245–390 (1937). While most of the experimental data and procedures have, of course, since been improved (e.g., see K. T. Bainbridge, in E. Segré, editor, **Experimental Nuclear Physics**, Vol. I), this historically valuable article gives full references and, for example, comparisons of calculated and experimental Q values.

For earliest good application of SRT to data on nuclear disintegration, see discussion of Cockcroft-Walton experiment by K. T. Bainbridge, "The Equivalence of Mass and Energy," Phys. Rev. 44, 123

(1933). See also Oliphant, Kinsey, and Rutherford, Proc. Roy. Soc. 141, 722–733 (1933).

62.A "Precision Measurement of the Wave-Length and Spectral Profile of Annihilation Radiation from Cu⁶⁴." J. W. M. DuMond, D. A. Lind, and B. B. Watson. Phys. Rev. 75, 1226–1239 (1949). As an example of a thorough experimental project; full description of method and equipment.

63.A "Summary of Recent Measurements of the Compton Effect." A. Bernstein and A. K. Mann. Am. J. Phys. 24, 445–450 (1956). Review of experimental work, with layouts and data.

Note: For experiments involving time dilation, see Sec. VII below.

VI. MORE ON INERTIA OF ENERGY

*64.E The Concept of Mass in Classical and Modern Physics. Max Jammer. (Harvard University Press, Cambridge, Massachusetts, 1961), 230 pp., Chapter 13. "Mass and Energy" is an excellent, fully documented review of the field from Maxwell on. Includes discussion of Einstein's 1905 paper (item 1) on inertia of energy, and its circular reasoning.

Note: For other articles by Einstein on mass-energy equivalence in SRT, see Ann. Physik 20, 627–633 (1906); 23, 371–384 (1907); Bull. Am. Math. Soc. 41, 223–230 (1935).

65.E "Derivation of the Mass-Energy Relation." H. E. Ives. J. Opt. Soc. Am. 42, 540–543 (1952). Intends to show that Einstein "did not derive the mass-energy relation" in 1905, and that the credit goes to Poincaré and Hasenöhrl. A better example of a small but vocal tradition.

*66.E The Theory of the Relativity of Motion. R. C. Tolman. (University of California Press, Berkeley, 1917), 225 pp. Derivation of E=mc² by collision of particles, pp. 35–41. Originated in article by Lewis and Tolman, Phil. Mag. 18, 510–523 (1909). Essentially the same treatment is in R. C. Tolman, Relativity, Thermodynamics and Cosmology, pp. 43–50, and many recent books.

*67.E Introduction to Atomic and Nuclear Physics. Otto Oldenberg. (McGraw-Hill Book Company, Inc., New York, 1961), 3rd ed., 380 pp. Appendix 7, pp. 362–364, has the photon-in-a-box derivation of E=mc², based on Max Born (item 4), and should be read together with a critique and extension of the method by Eugene Feenberg, "The Inertia of Energy." Am. J. Phys. 28, 565–566, 1960.

68.I "Masse und Energie in der speziellen Relativitätstheorie." R. Lämmel. Helv. Phys. Acta 12, 511–518 (1939). Uses Minkowskian representation for motion of c. of m. of two spheres to derive usual result (without using velocity addition theorem or energy conservation theorem).

69.E "L'Inertie de l'énergie et ses conséquences." P. Langevin. J. phys. radium 3, 553 ff. (1913). Reprinted in Oeuvres scientifiques de Paul Langevin (CNRS, Paris, 1950), pp. 397–426. Review article

with many examples. The same collection contains also reprints of two other review articles by Langevin in his characteristically clear and graceful style.

70.I "Zur Theorie der Raketen." J. Ackeret. Helv. Phys. Acta 19, 103–112 (1946). Relativistic motion of rocket with changing rest mass. A thorough and useful review, with examples. On the same general problem, see "Relativistic Mechanics of a Material Point of Variable Mass." N. S. Kalitsin, Soviet Phys. —JEPT 1, 565–567 (1955). Minkowskian analysis.

71.I "Special Relativity and the Electron." W. W. Harman. Proc. IRE 37, 1308–1314 (1949). A straightforward discussion of the modification of laws and techniques in engineering practice involving electrons at high speed.

72.I "Wave and Inertial Properties of Matter." Richard Schlegel. Am. J. Phys. 22, 77–82 (1954). Inertial properties of matter and mass-energy equivalence are derived from de Broglie equation, LT, and E=hν.

73.A "Relativistic Mechanics of a Particle." David Park. Am. J. Phys. 27, 311–313 (1959). Axiomatic development, to parallel Newtonian mechanics.

VII. TIME DILATION AND CLOCK PROBLEMS

From the earliest days of SRT, much attention has been paid to time dilation and the clock problem, or twin "paradox," or "voyageur de Langevin." By now, far too much seems to have been written on this. But students are intrigued by it as much as ever, and if the discussion is kept in bounds, it can serve a useful function.

74.I "Experimental Study of the Rate of a Moving Clock." H. E. Ives and G. R. Stilwell. J. Opt. Soc. Am. 28, 215–226 (1938). The basic paper includes clear photos and drawings. See also criticism by R. C. Jones, J. Opt. Soc. Am. 29, 337–339 (1939); reply by Ives and Stilwell, J. Opt. Soc. Am. 31, 369–374 (1941); repetition of original experiment by G. Otting, Physik Z. 40, 681–687 (1939); and summary of history of idea of rate dependence on speed, by H. E. Ives, J. Opt. Soc. Am. 37, 810–813 (1947).

*75.I "Variation of the Rate of Decay of Mesotrons with Momentum." Bruno Rossi and D. B. Hall. Phys. Rev. 59, 223–228 (1941). Experimental check of time dilation via the dependence of lifetime of cosmic ray mesons on their energy. For artificially produced mesons, an early check was by R. P. Durbin, H. H. Loar, and W. W. Havens, Jr., "The Lifetimes of the π⁺ and π⁻ Mesons," Phys. Rev. 88, 179–183 (1952). See also summary by F. S. Crawford, Jr. (item 79).

76.I "Some Recent Experimental Tests of the Clock Paradox." C. W. Sherwin. Phys. Rev. 120, 17–21 (1960). Discusses work of Pound-Rebka and of Hay-Schiffer-Cranshaw-Egelstaff as experimental verification of velocity dependence of clock rates.

*77.E "The Space Traveller's Youth." H. Bondi. Discovery 18, 505–510 (1957). For general reader, but with worked-out examples and Minkowskian diagrams. Elaborated by M. Born, Phys. Blätt. 14, 207–212 (1958).

*78.E "The Clock Paradox." G. BUILDER. Australian J. Phys. 10, 226–245 (1957). Uses only SRT, avoids the "paradox" by proper choice of method of calculating retardation. See also G. Builder, "The Resolution of the Clock Paradox," Phil. Sci. 26, 135–144 (1959).

*79.E "Experimental Verification of the 'Clock Paradox' of Relativity." F. S. CRAWFORD, JR. Nature 179, 35–36 (1957). Summarizes experiments on mesons by Rossi, Hilbert, and Hoag (1940), F. Rasetti (1941), P. M. S. Blackett (1937), and H. Ticho (1947), which, together, verify time dilation. (Reply by H. Dingle, Nature 179, 865–866 (1957) is followed by rejoinder by F. S. Crawford, Nature 179, 1071–1072 (1957).) Best read together with item 84.

*80.E "The Clock Paradox in Relativity." C. G. DARWIN. Nature 180, 976–977 (1957). A very simple discussion, showing asymmetry in logs kept by two space ships counting each other's regularly sent light flashes.

81.E "Time Dilation and Doppler Effect." J. D. ROBINSON AND E. FEENBERG. Am. J. Phys. 25, 490 (1957). Uses relativistic formula for longitudinal Doppler effect. See also E. Feenberg, Am. J. Phys. 27, 290 (1959).

82.E "Twin Paradox in Special Relativity." ROBERT H. ROMER. Am. J. Phys. 27, 131–135 (1959). Uses only simplest concepts of SRT, small velocities, and virtually no mathematics. Discusses some of the usual objections.

*83.E "Relativity and Space Travel." J. R. PIERCE. Proc. IRE 47, 1053–1061 (1959). Discusses quantitatively a clock paradox, a twin problem, frequency shift in gravitational field, clock rate on satellite, speed allowable by photon rocket, and insufficiency of interstellar matter to power a space ship. See also C. Darwin, item 80.

*84.I "The 'Clock Paradox' and Space Travel." E. M. MCMILLAN. Science 126, 381–384 (1957). Uses SRT and ordinary case as well as continuously accelerated coordinate system. Examples to show that "relativistic time modifications are negligible for travel within solar system" even for large accelerations, and that necessary energy expenditures to get important effect is "far beyond any foreseeable practical limits."

*85.I "Relativistic Observations and the Clock Problem." J. TERRELL. Nuovo cimento 16, 457–468 (1960). Perhaps the best of the recent reviews. Shows from several points of view where asymmetry lies. Only SRT involved. Good bibliography.

*86.I "On Solutions of the Clock Paradox." G. D. SCOTT. Am. J. Phys. 27, 580–584 (1959). A useful, brief survey of discussions via length contraction, Doppler effect, and world lines in chronogeometry; repeats that GRT adds no physically new aspect.

87.I "The Clock Paradox." C. C. MACDUFFEE. Science 129, 1359 (1959). Time dilation treated in two-

dimensional Minkowski space. Made even simpler in M. L. Boas, Science 130, 1471–1472 (1959).

88.I "Certain Matters in Relation to the Restricted Theory of Relativity, with Special Reference to the Clock Paradox and the Paradox of the Identical Twins." W. F. G. SWANN. Am. J. Phys. 28, 55–64, 319–323 (1960). Part I is a thoughtful study of "Fundamentals," Part II on the clock paradox.

89.I Electromagnetism and Relativity. E. G. CULLWICK. (Longmans Green and Company, London and New York, 1957), 299 pp. Chapter 5, pp. 62–80 and Appendix 2 on clock paradox, partial to Dingle's view.

90.A "Zum Uhrenparadoxon." M. BORN AND W. BIEM. Proc. Koninkl. Ned. Acad. Wetenschap. B61, 110–120 (1957). Approach via GRT and by simplification of solution in C. Møller, Theory of Relativity (1952). Another GRT treatment, generally considered also among the best, is by R. C. Tolman, Relativity, Thermodynamics, and Cosmology (1934), pp. 192–197.

Note: Several of the books cited above have good treatments of time dilation and clock problem, e.g., Arzeliès (item 2), Born (item 4), Sherwin (item 25), Stephenson and Kilmister (item 11), and Pauli (item 8).

There has been recently a well-known debate on the twin "paradox." For those who are interested in the points exchanged among the main discussants, this fairly complete chronological listing will be helpful: W. H. McCrea, Nature 167, 680 (1951); G. Thomson, *The Foreseeable Future* (Cambridge University Press, New York, 1955), pp. 88–89; H. Dingle, Nature 177, 782–784, 785 (1956); W. H. McCrea, Nature 177, 784–785 (1956); H. Dingle, Nature 178, 680–681 (1956); W. H. McCrea, Nature 178, 681–682 (1956); H. Dingle, Proc. Phys. Soc. (London) A69, 925–934 (1956); W. H. McCrea, Proc. Phys. Soc. (London) A69, 935 (1956); F. S. Crawford, Jr., Nature 179, 35–36 (1957); H. Dingle, Nature 179, 865–866 (1957); W. H. McCrea, Nature 179, 909–910 (1957); F. S. Crawford, Jr., Nature 179, 1071–1072 (1957); H. Dingle, Nature 179, 1242–1243 (1957); J. H. Fremlin, Nature 180, 499–500 (1957); H. Dingle, Nature 180, 500 (1957); E. M. McMillan, Science 126, 381–384 (1957); G. Builder, Australian J. Phys. 10, 226–245 (1957); G. Builder, Australian J. Phys. 10, 424–428 (1957); C. Darwin, Nature 180, 976–977 (1957); L. Essen, Nature 180, 1061–1062 (1957); H. Dingle, Nature 180, 1275–1276 (1957); R. Fisher and W. H McCrea, Discovery 18, 56–58 (1957); Lord Halsbury et al., Discovery 18, 174–175 (1957); H. Bondi, Discovery 18, 505–510 (1957); R. M. Frye and V. M. Brigham, Am. J. Phys. 25, 553–555 (1957); H. Dingle, Science 127, 158–160 (1958); and E. M. McMillan, Science 127, 160–162 (1958).

Since then the papers on this subject have tended less to be a debate involving Dingle; further bibliography to 1960 is given in footnotes 16 to 23 of item 85 (Terrell).

There also was an earlier exchange of articles on the concept of time among H. Dingle, P. S. Epstein, and L. Infeld, in Am. J. Phys. 10 (1942) and 11 (1943).

VIII. SOME OTHER CONSEQUENCES OF LT (E.G., VISUAL APPEARANCE)

*91.I "Invisibility of the Lorentz Contraction." JAMES TERRELL. Phys. Rev. 116, 1041–1045 (1959). It is shown—not without embarrassment to some of us—that to a single observer objects in rapid motion at a relatively large distance appear to have undergone rotation, not contraction. See also earlier paper for case of spheres by R. Penrose, Proc. Cambridge Phil. Soc. 55, 137–139 (1959).

*92.E "The Visual Appearance of Rapidly Moving Objects." V. F. Weisskopf. Phys. Today 13, 24–27 (1960). A beautiful summary of the work of J. Terrell (see item 91) and its implications.

93.E "Observation of Length by a Single Observer." Roy Weinstein. Am. J. Phys. 28, 607–610 (1960). On the length seen by a single observer: close to the trajectory of a rod the latter can appear elongated. However, for display of contraction by use of pulsed radar system, see C. W. Sherwin, Am. J. Phys. 29, 67–69 (1961). For case of nearby spheres and rods, see "Apparent Shape of Large Objects at Relativistic Speeds," Mary L. Boas, Am. J. Phys. 29, 283–286 (1961). For early work on appearance of objects, see also "The FitzGerald-Lorentz Contraction—Some Paradoxes and Their Resolution," W. H. McCrea, Sci. Proc. Roy. Dublin Soc. 26, 27–36 (1952). For extension of argument to measurement of time intervals by single observer, see A. D. Crowell, Am. J. Phys. 29, 370–371 (1961).

94.I "Note on the Stress Effects Due to Relativistic Contraction." E. Dewan and M. Beran. Am. J. Phys. 27, 517–518 (1959). A nice teaser. Will a thread between two identical rockets break as they leave at identical velocity, one behind the other? See also discussion of this article by A. A. Evett and R. W. Wangsness, Am. J. Phys. 28, 566 (1960). Another puzzle (will a length-contracted object fall through a grid over which it slides?) is in W. Rindler, Am. J. Phys. 29, 365–366, 859 (1961).

95.I "Relativity of Moving Circuits and Magnets." David L. Webster. Am. J. Phys. 29, 262–268 (1961). A necessary reminder that even for slow motions, the LT giving charge density cannot be approximated nonrelativistically. Illustrative examples for rotating and nonrotating circuits and magnets.

96.A "Lorentz Transformation and the Thomas Precession." W. H. Furry. Am. J. Phys. 23, 517–525 (1955). Kinematic effects of SRT treated by considering iterations of the infinitesimal LT.

IX. OTHER TEACHING AIDS

A. Laboratory and Demonstration Experiments

Almost nothing directly applicable to teaching SRT is now available, and very little that is indirectly applicable. But there is hope for early help with apparatus.

97. Speed of Light. Rotating mirror method, apparatus by Leybold Nachfolger, distributed in the U. S. by J. Klinger and others. Model 47640, $125 (plus $85 for optional attachments 31109, 46012a, 46320). For Kerr Cell shutter method, see C. H. Palmer, Jr., and G. S. Spratt, "A Laboratory Experiment on the Velocity of Light," Am. J. Phys. 22, 481–485 (1954). See also W. P. Alford and A. Gold, Am. J. Phys. 26, 481–484 (1958).

98. Michelson Interferometer. See T. B. Brown, editor, The Lloyd William Taylor Manual of Advanced Undergraduate Experiments in Physics (Addison-Wesley Publishing Company, Inc., Reading, Massachusetts, 1959), pp. 227–230. A commercial model ("M4 Interferometer") is sold by Atomic Laboratories, Inc., Berkeley, California. Model 87481–1 with Michelson Optics, $250; Model 87481–2 with Fabry-Perot Optics, $275; Model 87481–3 with both Optics, $395; Model 71859–26 Mercury Light Source, $15.

An inexpensive Michelson Interferometer is described by E. C. Cave and L. V. Holroyd, Am. J. Phys. 23, 61–63 (1955).

For microwave version of Michelson interferometer, see T. B. Brown, manual listed above, pp. 298–299; also T. G. Bullen, Am. J. Phys. 24, 525–526 (1956). A 3-cm microwave apparatus is available from Welch Scientific Company, Chicago, Model 2640, $385.

An inexpensive optical interferometer experiment is under design at the M.I.T. Science Teaching Center (directed by F. L. Friedman), and another at Educational Services Incorporated (E.S.I., directed by U. Haber-Schaim), where there is also work on a first-order (null) ether-drift experiment and an acoustical Fizeau experiment analogue.

99. Electron Dynamics at High Speeds. See M. Peter, "Lecture Demonstration of Relativistic Behavior of Electrons," Am. J. Phys. 23, 515–517 (1955).

V. Neher at California Institute of Technology is developing a Busch tube experiment for student laboratory. A number of projects are under way at the Science Teaching Center of M.I.T., involving Professors Frisch, Bertozzi, King, and Smith; they include experiments on limiting velocity of electrons in linear accelerator, time-of-flight measurement of electron beam, and a simplified Bucherer-Kaufmann apparatus. Some experiments are to be on film; others also are intended to yield apparatus for students.

The use of beta-ray spectrometers for student laboratories is described in T. B. Brown (item 98), pp. 499–505; by C. M. Fowler and R. Dean Dragsdorf, Am. J. Phys. 23, 108–210 (1955); in Experiment A-6, Supplement to Analytical Laboratory Physics, by H. F. Meiners, W. Eppenstein, K. H. Moore, and J. P. Nickol, Rensselaer Polytechnic Institute.

A commercial beta-ray spectrometer is sold by Atomic Laboratories, Inc., Model 79652, $295; it requires also Electromagnet Model 79641, $295, and scaler, PSU, vacuum pump, and gauge.

B. Films

Here again, virtually nothing is yet on the market specifically for teaching SRT. The nearest is the 26-minute film "Frames of Reference" (1958), produced by PSSC-ESI and distributed by Modern Talking Picture Service, Inc., 3 East 54 Street, New York 22. New York. Two filmed 50-minute lectures on SRT by G. Gamow are soon to be released by General Dynamics–Convair, San Diego, California.

SPECIAL RELATIVITY THEORY

SELECTED REPRINTS

A Project of the AAPT Committee on Resource Letters, 1963

CONTENTS

Reprint books in this series are available at $2. 00 each, or any combination
of 3 for $5. 00. Single orders to one address for any combination of 15 or
more reprint books, $1. 50 per copy from: American Institute of Physics,
335 East 45 Street, New York 17, New York   10017

On the Origins of the Special Theory of Relativity*

GERALD HOLTON

Department of Physics, Harvard University

(Received May 9, 1960)

Einstein's early work on relativity theory is found to be related to his other work at that time (e.g., in subject matter and style). In addition to this element of internal continuity one finds also—as a key to a new evaluation of the significance of Einstein's contribution—an external continuity with the classic, Newtonian tradition governing restrictions on permissible hypotheses. On the other hand, Einstein's work is shown to have been, in important respects, more independent of other contemporary work in this field than has recently been proposed.

These continuities and discontinuities are set forth to make the point that philosophical studies of scientific work should proceed on historically valid ground. Some guiding principles are indicated for dealing with conflicting source materials for such studies.

WHEN I received the persuasive invitation to speak today on a problem of theory construction and of the logic of discovery, I noted particularly the request to bring out the historical-sociological aspects. This directive was a pleasant surprise, for I recalled that Hans Reichenbach had flatly declared himself for the opposite view when he said "The philosopher of science is not much interested in the thought processes which lead to scientific discoveries···, that is, he is not interested in the context of discovery, but in the context of justification".[1] If, therefore, I shall make some remarks on the origins of Einstein's special theory of relativity, I will be disobeying the Reichenbachian dictum. However, I draw further strength for this resolution from Einstein, who himself declared for the value of the historical treatment of the rise of key theories in science.

In fact, it is appropriate to say at the very outset to an audience consisting primarily of philosophers of science that sound historical investigations have lately perhaps been overlooked as important bases of sound philosophical discussions. Some examples come to mind immediately. The crux of the Copernican revolution was initially not, as is maintained in some philosophical works, a pragmatic search for the smallest number of components with which to build a world system, nor was it the establishment of the possibility of relativism in the choice of coordinate systems. Rather, as historians of science have shown, it was a return to an earlier, even an Aristotelian austerity concerning the type of motion judged to be suitable for the construction of the world system, mixed with a commitment to a neo-Platonic epistemology that looked for the warrant of reality in a new direction. The importance of Kepler is not that he was a mystic, an obsessed searcher for empirical rules, or a master of the intuitive, "personal" way to scientific knowledge; on the contrary, it can be shown that he was the first of the modern mathematical physicists, the first to look with some success for one dynamical explanation of all celestial and terrestial motions. Galileo, we have had to relearn only recently, was not the patron saint of laboratory experimentation, as philosophers of science have at times maintained. Concerning the abuse to which Newton's work has had to lend itself, the less said the better.

Einstein's work has not been immune from this fate. I am suggesting that in this case, as in the others, we build our philosophical analyses of science on real ground instead of dubious models, that we examine what physics was like in Einstein's time, what he did and said, how he came to do and say these things, and how he changed his mind—not once, but often. I urge this not as an easy program—for it is not that—and even not just because it is in principle better to do justice to the work of a man on his own terms rather than to use his work for a purpose which

* Presented at the Symposium *Theory Construction in Logical and Historical Perspective* on December 27, 1959, organized by Section L (History and Philosophy of Science) of the AAAS, the American Philosophical Association, and the Philosophy of Science Association. Based on work-in-progress, supported in part by a grant of the National Science Foundation.

[1] P. A. Schilpp, editor, *Albert Einstein: Philosopher-Scientist* (Library of Living Philosophers, 1949), p. 292.

Reprinted from Am. J. Phys.
(Oct. 1960 - Pages 627-636)
By Permission

may have been inherently foreign to him. I urge this, rather, because I believe that a future source of strength of scholarship in the philosophy of science lies in philosophical analysis of historical cases.

I speak of Einstein's work because his case is both typical and special. The rise of relativity theory shares many features with the rise of other important scientific theories in our time, and in addition it is of course very much more: To find another work that illuminates as richly the relationship between physics, mathematics, and epistemology, or between experiment and theory, or one with the same range of scientific, philosophical and general intellectual implications, one would have to go back to Newton's *Principia*. The theory of relativity was a key development, both in physical science itself and also in modern philosophy of science. The reason for its dual significances is that Einstein's work provided not only a new principle of physics, but, as A. N. Whitehead said, "a principle, a procedure, and an explanation." Accordingly, the commentaries on the historical origins of the theory of relativity have tended to fall into two classes, each having distinguished proponents: the one views it as a mutant, a sharp break with respect to the work of the immediate predecessors of Einstein; the other regards it as an elaboration of then current work, e.g., by Lorentz and Poincaré.

To my mind, the Einsteinian innovation is understood best by superposition of both views, by seeing the discontinuity of methodological orientation within an historically continuous scientific development.[2] Before we come to discuss this, and if we take seriously my point of view, we should first be ready to investigate a number of real problems of the historical or even "historical-sociological" kind: What are the sources for a study of the origins of the special theory of relativity (*RT*) and what is their probable reliability? What was the state of science around 1905, what were the contributions which prepared the field for the *RT*, and what did Einstein know about them? What were the steps by which Einstein reached the conclusions he published in 1905? To what extent was this work a member of

a continuous chain having as its immediate predecessors Lorentz and Poincaré? What was the role of experiment in the genesis of the *RT*, and what the role of the existence of contradictory hypotheses? What part played epistemological analysis in Einstein's thought? What was the early reception of the *RT* among scientists? In particular, what was Einstein's relation with Mach, Lorentz, and Planck? What may we say about the style of Einstein's work and his personal orientations? What, if anything, in the origins and content of the *RT* is typical of other theories with great impact on science? And even, what methodological principles for the study of the history of science emerge from this study?

We would find that the existing literature is not always of help in studying such questions. The literature on the *RT* is of course vast. LeCat[3] listed over 3400 scientific papers in the field up to 1922, with an approximately exponential growth giving a sevenfold increase in seven years. Biographically or philosophically oriented analyses are also fairly numerous (for example, by Schlick, Reichenbach, Frank, Meyerson, Cassirer, Whitehead, Wenzel, Grünbaum, Polanyi, Margenau, Lenzen, Bridgman, and Northrop.) It may be remarked there has so far been no full-scale historical study (although one is now in progress). A number of valuable essays exist in this direction (for example, by Born, Dugas, Kuznetsov, von Laue, Pauli, Straneo, and Whittaker); these are generally concerned with the chronological development of physics, and typically constitute a portion of a longer work having a purpose different from that of a primarily historical-philosophical study. For the latter, the best source is at present indeed Einstein's own set of papers.

CONTINUITY IN EINSTEIN'S WORK

To these papers we must turn to discover, for example, the elements of continuity linking Einstein's first publication on the *RT* with his other work at the time and with the older tradition itself. After the paper of 1905,[4] Einstein returned to the exposition of the *RT* several times, and each restatement is of interest. For instance, in

[2] G. Holton, *IX Congreso Internacional de Historia de las Ciencias, Guiones de las Communicaciones* (Barcelona-Madrid, 1959), Vol. II, p. 41.

[3] Maurice LeCat, *Bibliographie de la Relativité* (Bruxelles, 1924).

[4] A. Einstein, Ann. Physik **17**, 891 (1905).

his book *Über die spezielle und die allgemeine Relativitätstheorie*[5] he emphasized in his introduction that "the author has made the greatest effort to present the main ideas···on the whole in the sequence and in such context as they in fact arose." It is not surprising that the sequence given there is not in accord with the sequence of steps in the 1905 paper itself, but the historian of science finds an interesting problem in the fact that neither of these is in accord with other autobiographical or biographical accounts.

When one studies the relativity papers in the larger contextual setting of Einstein's other scientific papers, particularly those on the quantum theory of light and on Brownian motion which also were written and published in 1905, one notices two crucial points. While the three epochal papers of 1905—sent to the *Annalen der Physik* at intervals of less than eight weeks—seem to be in entirely different fields, closer study shows that they arose in fact from the same general problem, namely, the fluctuations in the pressure of radiation. In 1905, as Einstein later wrote to von Laue,[6] he had already known that Maxwell's theory leads to the wrong prediction of the motion of a delicately suspended mirror "in a Planckian radiation cavity." This connects on the one hand with the consideration of Brownian motion as well as to the quantum structure of radiation, and on the other hand with Einstein's more general reconsideration of "the electromagnetic foundations of physics" itself.[7]

One also finds that the style of the three papers is essentially the same, and reveals what is typical of Einstein's work at that time. Each begins with the statement of formal asymmetries or other incongruities of a predominantly esthetic nature (rather than, for example, a puzzle posed by unexplained experimental facts), then proposes a principle—preferably one of the generality of, say, the second law of thermodynamics, to cite Einstein's repeated analogy—which removes the asymmetries as one of the deduced consequences, and at the end produces one or more experimentally verifiable predictions.

Specifically, Einstein's first paper on the quan-

tum theory of light opens in a typical manner: "There exists a radical formal difference between the theoretical representations which physicists have constructed for themselves concerning gases and other ponderable bodies on the one hand, and Maxwell's theory of electromagnetic processes in so-called empty space on the other hand."[8] The significant starting point is a formalistic difference between theoretical representations in two fields of physics which, to most physicists, were so widely separated that no such comparison would have invited itself and therefore no such discrepancy would be noted. The discrepancy Einstein points out is between the discontinuous or discrete character of particles and of their energy on one hand, and the continuous nature of functions referring to electromagnetic events and of the energy per unit area in an expanding wave front on the other hand. The discussion of the photoelectric effect, for which this paper is mostly remembered, occurs toward the end, in a little over two pages out of the total sixteen. The prescription for obtaining an experimental verification of his point of view is given in a single, typically succinct Einsteinian sentence (straight-line relation with constant slope between frequency of light and stopping potential for all electrode materials).

In his second paper published in 1905,[9] Einstein points out in the second paragraph that the range of application of classical thermodynamics may be discontinuous even in volumes large enough to be microscopically observable. He ends with the equation giving Avogadro's number in terms of observables in the study of particle motion, and with the one-sentence exhortation: "May some investigator soon succeed in deciding the question which has been raised here, and which is important for the theory of heat!" Significantly, Einstein reported the following year[10] that after the publication of this paper was his attention drawn to the experimental identification, as long ago as 1888, of Brownian motion with the effect whose existence he had deduced as a necessity from the kinetic-molecular theory. In his autobiographical notes he repeats that he did the work of 1905 "without knowing that observa-

[5] A. Einstein, (Braunschweig, 1916).
[6] Letter of January 17, 1952 (unpublished). See also Max Born in *Fünfzig Jahre Relativitätstheorie*, edited by A. Mercier and M. Kervaire (Bern, 1955), pp. 248–249.
[7] See footnote reference 1, p. 47.

[8] A. Einstein, Ann. Physik **17**, 132 (1905).
[9] A. Einstein, Ann. Physik **17**, 549 (1905).
[10] A. Einstein, Ann. Physik **19**, 371 (1906).

tions concerning Brownian motion were already long familiar".[11]

The third paper of 1905[12] is, of course, Einstein's first paper on the RT. He begins again by drawing attention to a formal asymmetry, i.e., in the description of currents generated during relative motion between magnets and conductors. The paper does not invoke explicitly any of the several well-known experimental difficulties—and the Michelson and Michelson-Morley experiments are not even mentioned when the opportunity arises to show in what manner the RT accounts for them. At the end, Einstein briefly mentions here, too, specific predictions of possible experiments (giving the equation "according to which the electron must move in conformity with the theory presented here").[13]

RETURN TO A CLASSIC RESTRICTION ON HYPOTHESES

The recognition of these common elements in the three papers prepares us for the essential realization that the fundamental postulates appearing in each of the three papers are *heuristic*. The heuristic nature of the postulate of relativity was from the beginning apparent to Einstein (as he asserted in 1907 and later) because of the restriction of the RT to translational motions and to gravitation-free space.[14]

The study of the three papers together reveals also the extent to which Einstein's RT represents an attempt to restrict hypotheses to the most *general kind* and the *smallest number* possible—a goal on which Einstein often insisted.[15] In the 1905 paper on RT, he makes, in addition to the

[11] See footnote reference 1, p. 47. See also L. Infeld, *Albert Einstein* (New York, 1950), p. 97–98.
[12] A. Einstein, Ann. Physik 17, 891 (1905).
[13] See footnote reference 12, p. 921.
[14] On a few occasions, although not in the original paper, Einstein made this point [e.g., Ann. Physik 23, 206 (1907)]: "The relativity principle [is to be regarded]··· solely as a heuristic principle, which, considered by itself, contains only assertions about rigid bodies, clocks, and light signals."
[15] Cf. A. Einstein, "The Problem of Space, Ether, and the Field in Physics," *Ideas and Opinions by Albert Einstein*, translated and revised by Sonja Bargmann (New York, 1954), p. 282: "The theory of relativity is a fine example of the fundamental character of the modern development of theoretical science. The initial hypotheses become steadily more abstract and remote from experience. On the other hand, it gets nearer to the grand aim of all science, which is to cover the greatest possible number of empirical facts by logical deductions from the smallest possible number of hypotheses or axioms."

two "conjectures" raised to "postulates" (i.e., of relativity and of the constancy of light velocity) only four other hypotheses: one of the isotropy and homogeneity of space, the others concerning three logical properties of the definition of synchronization of watches. In contrast, H. A. Lorentz's great paper which appeared a year before Einstein's publication[16] and typified the best work in physics of its time—a paper which Lorentz declared to be based on "fundamental assumptions" rather than on "special hypotheses"—contained in fact eleven *ad hoc* hypotheses: restriction to small ratios of velocities v to light velocity c; postulation *a priori* of the transformation equations (rather than their derivation from other postulates); assumption of a stationary ether; assumption that the stationary electron is round; that its charge is uniformly distributed; that all mass is electromagnetic; that the moving electron changes one of its dimensions precisely in the ratio of $(1-v^2/c^2)^{\frac{1}{2}}$ to 1; that forces between uncharged particles and between a charged and uncharged particle have the same transformation properties as electrostatic forces in the electrostatic system; that all charges in atoms are in a certain number of separate "electrons"; that each of these is acted on only by others in the same atom; and that atoms in motion as a whole deform as electrons themselves do. It is for these reasons that Einstein later maintained that the RT grew out of the Maxwell-Lorentz theory of electrodynamics "as an amazingly simple summary and generalization of hypotheses which previously have been independent of one another····."[17]

If one has studied the development of scientific theories, one notes here a familiar theme: *the so-called scientific "revolution" turns out to be at bottom an effort to return to a classical purity.* This is not only a key to a new evaluation of Einstein's contribution, but indicates a fairly general characteristic of great scientific "revolutions." Indeed, while it is usually stressed that Einstein challenged Newtonian physics in fundamental

[16] H. A. Lorentz, Proc. Acad. Sci. Amsterdam 6, 809 (1904). This paper, originally presented as part of the proceedings of the meeting of April 23, 1904, was first published in June, 1904 in the Dutch language edition of the Proceedings [12, 986–1009 (1904)].
[17] See footnote reference 5, p. 28. See also A. Einstein, Scientia 15, 338 (1914).

ways, the equally correct but neglected point is
the number of methodological correspondences
with earlier classics, for example, with the
Principia.

Here a listing of some main parallels between
the two works must suffice: the early postulation
of general principles which in themselves do not
spring directly from experience; the limitation
to a few basic hypotheses[18]; the exceptional attention to epistemological rules in the body of a
scientific work; the philosophical eclecticism of
the author; his ability to dispense with mecha-
nistic models in a science which in each case was
dominated at the time by such models[19]; the
small number of specific experimental predictions; and the fact that the most gripping effect
of the work is its exhibition of a new point of
view.

The central problem, moreover, is the same in
both works: the nature of space and time, and
what follows from it for physics. Here, the basic
attitudes have in both cases more in common
than appears at first reading. That Newton's
absolute space and absolute time were not meaningful concepts in the sense of laboratory opera-
tions, was, of course, not the original discovery of
Mach; rather, it was freely acknowledged by
Newton himself. But Einstein was also quite explicit that in replacing absolute Newtonian space
and time with an infinite ensemble of rigid meter
sticks and ideal clocks he was not proposing a
laboratory-operational definition. He stated it
could be realized only to some degree, "not even
with arbitrary approximation," and that the
fundamental role of the whole conception, both
on factual and on logical grounds "can be at-
tacked with a certain right."[20] Thus the *RT*
merely shifted the locus of space time from the
sensorium of Newton's God to the sensorium of
Einstein's abstract *Gedanken*experimenter—as it
were, the final secularization of physics.

In his tribute on the occasion of the 200th
anniversary of Newton's death, Einstein wrote:
"I must emphasize that Newton himself was
better aware of the weakness inherent in his
intellectual edifice that the generation of learned
scientists which followed him. This fact has always aroused my deep admiration····."[21] He then
immediately draws attention to the fact that
"Newton's endeavors to represent his system as
necessarily conditioned by experience and to introduce the smallest number of concepts not
directly referable to empirical objects is everywhere evident." He recalls that Newton regarded
the law of gravitational interaction as a heuristic
device, "not supposed to be a final explanation,
but a rule derived by induction from experience."
When the essay ends with Einstein clearly as-
sociating himself with a view of causality which
he characterizes as "Newtonian," he could well
have widened the context of that remark.

TIME-DEPENDENCE IN SOURCE MATERIALS

I cannot avoid a word of warning on the use of
sources such as Einstein's writings, particularly
to an audience not professionally engaged in the
study of the history of science. This has to do
with the fact that in many important particulars
the writings of one man do not by any means
necessarily overlap. I am not speaking merely of
the fact that Einstein regarded the discoverer,
and particularly himself, as a very poor source of
information concerning the genesis of his own
ideas, and suggested rather that this study was
one of the most interesting tasks for the historian
of science. No, I have in mind the simple, yet
often neglected fact that Einstein as a person
with a single, unchanging identity, in a real sense
never existed, just as there never was a single
unchangeable entity called Galileo or Newton or
Dalton. Einstein himself saw this clearly when he
wrote at the start of his autobiographical notes[22]:

The exposition of that which is worthy of communication
does nonetheless not come easy; today's person of 67 is by
no means the same as was the one of 50, of 30, or of 20.
Every reminiscence is colored by today's being what it is,
and therefore by a deceptive point of view.... And it is not

---

[18] Wolfgang Pauli, in *Theory of Relativity* [(B. G. Teubner, Leipzig, 1921 and Pergamon Press, New York, 1958),
p. 5], unwittingly draws forceful attention to this particular point when summarizing his analysis of the *RT* in
the following words: "The postulate of relativity implies
that a uniform motion of the center of mass of the universe relative to a closed system will be without influence
on the phenomena in such a system." Note the correspondence with the main hypothesis in the last edition of the
Principia.
[19] Cf. Max von Laue, *Naturwissenschaften* **43**, 1 (1956).
[20] *Les Prix Nobel en 1921–1922* (Stockholm, 1923), p. 2.
See also A. Einstein, Naturwissenschaften 6, 692 (1918).

[21] A. Einstein, Naturwissenschaften **15**, (1927); reprinted in *Ideas and Opinions of Albert Einstein* (New York,
1954) 257.
[22] See footnote reference 1, pp. 3–7.

only growth or change—it is also the difference between experience lived and experience reported. In this case it is well possible that such an individual in retrospect sees a uniformly systematic development, whereas the actual experience takes place in kaleidoscopic particular situations.

These two effects, coupled with Einstein's large output of writings of both a scientific and a popular kind, explain why everyone—from the extreme positivist to the critical realist—can find some part of Einstein's work to nail to his mast as a battle flag against the others.

There are two ways of dealing with this problem in historically oriented work. The first is to be explicitly careful in the evaluation of all sources, including autobiographical statements, to allow a time-dependent weighting factor. This has always been true, but is particularly pertinent in modern physics where changes per unit time are far larger than before. Revealing examples, and very worthwhile topics of study, are Einstein's attitude to the ether problem, or his relation to Ernst Mach, or his more general epistemological position. Concerning the first of these, for instance, Einstein underwent a profound change of orientation between the statement near the beginning of his fundamental 1905 paper: "The introduction of a 'luminiferous ether' will prove to be superfluous inasmuch as the view here to be developed will not require an 'absolutely stationary space' provided with special properties"—a provocative remark on which Dugas astutely comments "such a declaration, made on the threshold of his theory, could only alienate him from the physicists imbued with the classical representation"[23]—to his Leiden speech of 1920 on "Äther und Relativitätstheorie" in which he says near the close: "Recapitulating, we may say that according to the general theory of relativity space is endowed with physical qualities; in this sense, therefore, there exists an ether. According to the general theory of relativity, space without ether is unthinkable; for in such a space there not only would be no propagation of light, but also no possibility of existence for standards of space and time (measuring rods and clocks), nor therefore any space-time intervals in the physical sense."[24]

[23] René Dugas, *A History of Mechanics* (New York, 1955), p. 490.
[24] In A. Einstein, *Sidelights on Relativity* (Methuen and Company, Ltd., London, 1922), p. 23. The next sentences

To the student of the nature of scientific theories, a sequence of individual documents on a particular topic from one pen represents therefore, as it were, a sequence of cross sections in space-time, from which he is challenged to reconstruct the progress or worldline of the topic. Particularly in recent and contemporary physics, no single segment of this worldline may be safely extrapolated; a quick turn is always likely. This enhances the interest: the reconstruction of the changing course of opinion on a topic becomes doubly important, and these changes in one topic may often be correlated with changes in another topic. In the case of Einstein, for example, the attitudes to the ether, to Mach, to epistemology and metaphysics generally, and to religion, all show closely correlated changes in time. This itself poses new and valuable problems, both to the historian and to the philosopher of science.

THE COMPLEMENTARITY OF SOURCE MATERIALS

There is a second problem involving divergent or contradictory views concerning a scientist's work. It is generated not by internal changes or conflicts, but by external ones. I can discuss this in the briefest way by pointing to the question of what one is to do with biographical works which are not in agreement.

Such biographies are a precious set of sources for the study of the origins of the relativity theory. Among the principal ones that appeared in Einstein's own time are, in order of publication, those by Moszkowski, Reiser, Reichinstein, Marinoff and Wayne, Seelig, Frank, Infeld, and Vallentin. Each has interest in its own right, but naturally enough they differ vastly in their points of view as well as on factual matters. One can begin to discern the Vivianis and Stukleys now, the sources of future myths and the sources of

reaffirm the difference between this and other ether models. "But this ether may not be thought of as endowed with the quality characteristic of ponderable media, as consisting of parts which may be tracked through time. The idea of motion may not be applied to it."
In this connection, see also the essay on "Relativity and the Problem of Space," which Einstein added in the 16th edition of *Relativity, the Special and General Theory* (Methuen and Company, Ltd., London, 1952). Commenting on it to Carl Seelig [*Albert Einstein* (Zürich, 1954), p. 291], Einstein wrote: "In particular, it is shown that the development has a close connection with Descartes' argument for the non-existence of 'empty space'."

reliable references. It was therefore important to discover the unpublicized fact that one of these was written under a pseudonym by a relative of Einstein and checked by him for factual accuracy, that another was publicly disowned by Einstein, that he made an attempt in a third case to persuade the author—whom he did not trust to be fair or accurate—to forgo publication, that he was pleased with the material in another of these books, and so forth.

The uncommonly large amount and variety of material emphasizes the problems the historian of science must face. The different points of view from which two or more honest biographies are written yield, of course, different interpretations. On some matters of "fact" (as, for example, dates and places) one certainly can ask for agreement or accuracy in some absolute sense. But on larger and more qualitative questions (for example, the acceptance of the theory) one can profitably adopt the attitude that evidence obtained by biographical research under different points of view cannot be comprehended within a single picture, but must be regarded as complementary in the sense that *only the totality of the presentations exhausts the possible information about the subject.* This will be recognized as closely analogous to one part of the complete statement of the complementarity principle in physics.[25] To look for an "independent" view in qualitative matters in any other way is likely to lead one to take merely a position equidistant between all others, or between the "isms" that motivate them.

The complementarity principle tells the physicist also that it is not possible to make a sharp separation between the behavior of atomic objects and the interaction with the measuring instruments which serve to define the conditions under which the phenomena appear. This statement, too, has a close parallel in the study of the history or philosophy of science, and one must therefore be aware that the scholar and the subject of his study together form one system in which it is not meaningful to try to achieve a complete separation of one part from the other. It is in this spirit that one must understand, and use, the picture of Einstein as a revolutionary which is painted by a revolutionary, and that of

Einstein as a positivist as presented by a positivist. To one who is committed to the existence of a real medium to explain the transmission of light through space, the *RT* is important primarily insofar as it adds to or subtracts from this position. Only with this explicit recognition can one use a number of accounts together, each of which would otherwise appear to present a strikingly different person or work.

WHITTAKER'S ACCOUNT OF THE ORIGINS OF EINSTEIN'S WORK

To illustrate this point concretely I wish to turn to a question on which a dispute has been active, namely, to what extent Einstein's work was original rather than anticipated by, or specifically based on, other published work. Particularly interesting is the essay on Einstein by Sir Edmund Whittaker in the *Biographical Memoirs of Fellows of the Royal Society* (London, 1955). Whittaker's commitment to the 19th-century tradition of physics and to the ether theory is illustrated in his well-known book *A History of the Theories of Aether and Electricity* up to about 1900 (London, 1910; 2nd ed., 1951), and also by his excellent contributions in the field of classical mechanics. Moreover, in the second volume of the *History*, completed in 1953, which carries the story to 1926, Whittaker had largely dismissed Einstein's paper of 1905 on the *RT* as one "which set forth the relativity theory of Poincaré and Lorentz with some amplifications, and which attracted much attention."[26]

This presentation evoked considerable criticism, some of which I know to have reached Whittaker while his book was still in manuscript, and some of which reached him by the time he composed the biographical memoir after Einstein's death in 1955. It is therefore noteworthy that in his 1955 necrolog for Einstein, Whittaker has not changed his earlier evaluation. For example, he repeats that Poincaré in a speech in St. Louis, U. S. A., in September 1904[27] had coined the phrase "principle of relativity." Whittaker asks how physics could have been reformu-

---

[25] I employ it here as a suggestive, though not prescriptive, analogy.

[26] Sir Edmund Whittaker, *A History of the Theories of Aether and Electricity: The Modern Theories 1900–1926* (New York, 1954), p. 40.
[27] J. H. Poincaré, *Bull. Sci. Math.* (1904); English translation in *Monist*, 15, 1 (1905).

lated in accordance with "Poincaré's principle of relativity," and he reports that with respect to the laws of the electromagnetic field this "discovery was made in 1903 by Lorentz," citing a paper by Lorentz in the *Proceedings of the Academy of Sciences, Amsterdam*, for the year 1903.[28] Whittaker shows that "the fundamental equations of the aether in empty space" are invariant under suitably chosen (i.e., Lorentz) transformations, and he concludes with the remarkable sentence: "Einstein in [the *RT* paper of 1905] adopted Poincaré's principle of relativity, using Poincaré's name for it, as a new basis for physics and showed that the group of Lorentz transformations provided a new analysis connecting the physics of bodies in motion relative to each other."[29]

Since Whittaker's analysis has been and is likely to continue to be given considerable weight, it is necessary to examine it closely. It turns out to be an excellent example of the proposition that no such analysis can be considered meaningful except insofar as it deals both with the material it purports to cover *and* with the prior commitments and prejudices of the scholar himself. Here is a brief summary of main findings when Whittaker's analysis is considered in this light.

(1) Einstein's *RT* paper of 1905 was indeed one of a number of contributions by many different authors in the general field of the electrodynamics of moving bodies. In the *Annalen der Physik* alone there are eight papers from 1902 up to 1905 concerned with this general problem. Einstein himself always insisted on this aspect of continuity. The earliest evidence is in a letter written in the spring of 1905 to his friend Conrad Habicht, describing his various investigations. In one sentence he describes the developing *RT* paper: "The fourth work lies at hand in concept [*liegt im Konzept vor*] and is an electrodynamics of moving bodies making use of a *modification* of the theory of space and time; you will surely be interested in the purely kinematic part of this work." (In Carl Seelig, see footnote reference 24, p. 89. Ital. suppl.) Seelig (*ibid.*, p. 97) also quotes a later remark of Einstein which gives in one

sentence his often repeated attitude: "With respect to the theory of relativity it is not at all a question of a revolutionary act, but of a natural development of a line which can be pursued through centuries."

On the other hand, to say that Einstein's paper "attracted much attention" is correct only if one neglects the first few years after publication. For the early period, a more characteristic reaction was, in fact, either total silence or the response to be found in the first paper in the *Annalen der Physik* that mentioned Einstein's work on the *RT*. It was a categorical experimental disproof of Einstein's theory by the eminent physicist W. Kaufmann, who concluded[30]:

"I anticipate right here the general result of the measurements to be described in the following: *The measurement results are not compatible with the Lorentz-Einsteinian fundamental assumption.*"

(2) The paper by Poincaré of 1904 which Whittaker cites turns out not to enunciate the new relativity principle, but is rather a very acute and penetrating though qualitative summary of the difficulties which contemporary physics was then making for six classical laws or principles, including what is in effect the Galilean-Newtonian principle of relativity. The list given by Poincaré is as follows: The law of conservation of energy; the second law of thermodynamics; the third law of Newton; "the principle of relativity, according to which the laws of physical phenomena should be the same whether for an observer fixed or for an observer carried along in a uniform movement or translation · · ·"; the principle of conservation of mass; and the principle of least action.[31] Of the principle of relativity Poincaré complains that it "is battered" by current developments in electromagnetic theory, although, he says, it "is confirmed by daily experience" and "is imposed in an irresistible way upon one's good sense." Poincaré's main point is to show the need for a new development, the outlines of which he suggests in these words: "Perhaps likewise we should construct a whole new mechanics, that we only succeed in catching a glimpse of, where inertia increasing with the velocity, the velocity of light would become an

[28] The citation given is "Proc. Acad. Sci. Amst. (English ed.) (1903) 6, 809."
[29] Sir Edmund Whittaker, in *Biographical Memoirs of Fellows of the Royal Society* (London, 1955), p. 42.

[30] W. Kaufmann, Ann. Physik 19, 495 (1906). Italics in original.
[31] See footnote reference 27, p. 5.

impassable limit."[32] Thus he illustrates both the power of his intuition and the qualitative nature of the suggestion.

(3) It is more difficult to discuss the 1903 paper of Lorentz which Whittaker, both in his book and in his Memoir, cited specifically as the work that spelled out most of the basic details of Einstein's *RT* of 1905. In the first place, this paper does not exist. What Whittaker clearly wished to refer to is the paper Lorentz published a year later, in 1904.[16] Since Whittaker was otherwise very careful with the voluminous citations of references, this repeated slip, which doubles the time interval between the work of Lorentz and of Einstein, is not merely a mistake. It is at least a symbolic mistake—symbolic of the way a biographer's preconceptions interact with his material.

(4) Whittaker clearly implied that Einstein used Lorentz's transformation equation published in 1904. He therefore chose to neglect that both Einstein and those close to him have repeatedly said that Einstein had not read Lorentz's 1904 paper.[33]

(5) Even if one does not wish to rely on the word of Einstein and other prominent physicists of his time in this matter, there are four items of internal evidence in Einstein's 1905 paper which indicate that he had not read Lorentz's of 1904. Einstein does write the transformation equations in à form equivalent to those of Lorentz (or, for that matter, of Voigt's of 1887); but whereas Lorentz had assumed these equations *a priori* in order to obtain the invariance of Maxwell's equations in free space, Einstein *derived* them from the two fundamental postulates of the *RT*. He therefore did not need to know of Lorentz's paper of 1904.[34]

Secondly, as Einstein's first two major papers of 1905 show, he was in the habit of giving credit in footnotes to the work of others which he might be using; the absence of a specific reference to the 1904 paper of Lorentz may therefore be taken at its face value, the more so since Einstein twice in the text of this same paper refers to Lorentz by name in citing the then current electromagnetic theory in the form Lorentz had given it in his book of 1895.[35] Parenthetically one may also say that it is rather preposterous to suggest that a young man of Einstein's temperament and painful honesty, and one who, as the letters to Lorentz soon thereafter show, revered Lorentz deeply, should knowingly be using, without acknowledgment, an important new finding in the recent work of the foremost theoretical physicist in this field.[36]

Next, in the second paragraph of his paper, Einstein recalls that the "laws of electrodynamics and optics" have been found to "be valid for all frames of reference for which the equations of mechanics hold good" to the first order of the quantity v/c. But one of the main points of the 1904 paper of Lorentz was his claim to have extended the theory to the *second* order of v/c. And a fourth internal evidence is Einstein's choice of convention in the expression for force and mass in the dynamics of charged particles; this choice[37] is far less suitable than Lorentz's, forcing Planck to point this out in 1906.

(6) Quite apart from the question whether Einstein's 1905 paper was written independently

---

[32] See footnote reference 27, p. 23.

[33] See the footnote on this point by A. Sommerfeld in the reprints and translations of Einstein's 1905 paper in the Teubner and Methuen editions of the collection of essays on the *RT* [e.g., *The Principle of Relativity* (London, 1923)]; or Pauli (footnote reference 18, p. 3); or Einstein's letter to Carl Seelig: "As for me, I knew only Lorentz's important work of 1895···but not Lorentz's later works and also not the inquiries of Poincaré connected with them. In this sense my work of 1905 was independent." [Techn. Rundschau, Bern, **47**, (May 6, 1955); cited in Max Born, footnote reference 6, p. 248.]

[34] This is by no means the only such case in Einstein's early scientific career. In fact, his work on thermodynamics and fluctuation phenomena in the period 1902–1905 was to a large extent a repetition of available material; as Einstein said later, "Not acquainted with the earlier investigations of Boltzmann and Gibbs, which had appeared earlier and actually exhausted the subject, I developed the statistical mechanics and the molecular-kinetic theory of thermodynamics which was based on the former." (P. A. Schlipp, footnote reference 1, p. 47). Einstein's unawareness in 1905 of the earlier identification of Brownian motion has been referred to previously. Anton Reiser provides the report [in *Albert Einstein* (New York, 1930), p. 52] that at his university Einstein planned to build a device for measuring the ether drift, not knowing of Michelson's apparatus; although this earliest example is quite understandable in terms of the incompleteness of Einstein's training at that point, it illustrates a remark made often about him by his friends: that he read little, but thought much.

[35] H. A. Lorentz, *Versuch einer Theorie der elektrischen und optischen Erscheinungen in bewegten Körpen* (Leiden, 1895).

[36] Einstein later accurately reported that "At the turn of the century, H. A. Lorentz was regarded by theoretical physicists of all nations as the leading spirit; and this with fullest justification." A. Einstein, in *H. A. Lorentz*, edited by G. L. de Haas-Lorentz (Amsterdam, 1957), p. 5.

[37] As most recently remarked by Max von Laue [Naturwissenschaften **43**, 4 (1956)] in documenting his belief that Einstein did not know of Lorentz's 1904 paper.

of Lorentz's is the equally significant fact that in a crucial sense Lorentz's paper was of course not on the relativity theory as we understand the term since Einstein. Lorentz's fundamental assumptions are not relativistic; as Born says, "he never claimed to be the author of the principle of relativity,"[38] and, on the contrary, referred to it as "Einstein's Relativitätsprinzip" in his lectures of 1910. In Lorentz's essay on "The Principle of Relativity of Uniform Translation," published in 1922,[39] six years before Lorentz's death, he still asked that space be considered to have "a certain substantiality; and if so, one may, in all modesty, call true time the time measured by clocks which are fixed in this medium, and consider simultaneity as a primary concept."[40] In his 1904 paper he had postulated the nonrelativistic addition theorem for velocities, $v = V + u$, and even in the 1922 book he did not consider the velocity of light as inherently the highest attainable velocity of material bodies.

(7) Finally, we note another set of important differences between Lorentz's accomplishment of 1904 and what Whittaker implies. Strictly speaking, the Lorentz theory of 1904 applies only to small values of v/c, since the constant l which is taken to be 1 for small values of v/c enters in the first power in the transformation equations for x and t. Also, Maxwell's equations in the presence of charges are not completely invariant in Lorentz's treatment even at small speeds v, since in the primed (moving) system, a term is left over in the expression for $\mathrm{div}'D'$, namely, $\mathrm{div}'D' = [1 - (vu_x'/c^2)]\rho'$, as compared to $\mathrm{div}D = \rho$.[41] We have already noted the number of $ad\ hoc$ hypotheses which Lorentz was forced to introduce, and which, Einstein felt, robbed the theory of electromagnetic phenomena of the generality typical of fundamental conceptions.

In closing, I return to my initial remarks: The detailed study of the historical situation is, to my mind, an important first step in those discussions which try to base epistemological considerations on "real" cases. This is not always done easily; but it is through the dispassionate examination of historically valid cases that we can best become aware of the preconceptions which underlie all philosophical study.

---

[38] Max Born, footnote reference 6, p. 247.

[39] A. D. Fokker, editor, translated as Vol. III of *Lectures on Theoretical Physics* (London, 1931).

[40] See footnote reference 39, p. 211.

[41] Whittaker (footnote reference 26, p. 31), says Lorentz "obtained a transformation in a form which is exact to all orders of the small quantity v/c," although strictly speaking this is correct only for free space and relatively small values of v.

MASS-ENERGY RELATION

In modern physics: Electron mass dependent on velocity; mass and energy interchangeable. Energy of 1 gm mass equivalent to the total ultimate energy output of Boulder Dam during a 19-hour period

By DR. SAUL DUSHMAN

Assistant Director, Research Laboratory, General Electric Company

IN ORDINARY (or classical) mechanics we have been accustomed to assume that mass is independent of velocity. That this assumption might not be valid was recognized by Newton, in his definition of force as the rate of change of momentum

$$F = d \ (mv)/dt \qquad (1)$$

Assuming that mass is independent of velocity, this equation is replaced in classical mechanics by the relation

$$F = mdv/dt = m \ (d^2s/dt^2) = m \ \alpha \qquad (2)$$

where s is the distance traversed at the end of time t' and α denotes the acceleration. This leads to the further definition of kinetic energy

$$K = \frac{1}{2} \ mv^2 \qquad (3)$$

Mass of Electron Dependent on Velocity

After J. J. Thomson demonstrated the existence of the electron and Rutherford and others observed that radioactive atoms emit electrons with speeds approaching that of light (beta particles), speculation on the origin of the mass of the electron led to the view that this mass should vary with the speed. An electron in motion establishes a magnetic field. Energy is required to establish this field. Might it not be possible that the entire mass of the electron is electromagnetic?

A number of physicists attempted to carry out calculations on this point. M. Abraham[1] derived a relation for the increase in mass with velocity which has since been shown to be in disagreement with observations. Following this investigation, H. A. Bucherer,[2] and also H. A. Lorentz[3] deduced a relation for the electromagnetic mass, of the form

$$m = m_0 / \sqrt{1 - \beta^2} \qquad (4)$$

where m = mass of particle in motion
m_0 = rest mass
and $\beta = v/c$ = ratio of velocity to that of light

A. Einstein[4] showed in 1905 that the assumptions made by Lorentz are quite unnecessary since Equation (4) is a logical consequence from the principle of relativity, without any need of assuming that the mass of the electron is of electromagnetic origin.

(1)Ann. d. Physik, 10, 105 (1903).
(2)Math. Einführung in die Elektronentheorie, (1904).
(3)Theory of Electrons, (1909).
(4)Ann. d. Physik, 17, 891 (1905).

This equation states that the mass of a particle increases with the velocity. While the increase for ordinary velocities is extremely small, it becomes extremely large as the velocity approaches that of light until finally m/m_0 tends to become infinitely large. This rapid increase for large values of β is indicated in Table I.

TABLE I

$\beta = v/c$	$1 - \beta$	m/m_0
0.01	0.99	1.0005
0.1	0.90	1.010
0.25	0.75	1.033
0.50	0.50	1.154
0.75	0.25	1.512
0.90	0.10	2.294
0.95	0.05	3.203
0.99	0.01	7.087
0.995	0.005	10.013
0.999	10^{-3}	22.366
	10^{-4}	70.712
	10^{-5}	223.61
	10^{-6}	707.11

For values of β that are approximately equal to 1, it is evident that

$$\frac{m}{m_0} = 1/\sqrt{2 \ (1 - \beta)} = 0.7071/\sqrt{1 - \beta}$$

As a result of all this discussion on the increase of mass with velocity, a number of experimenters carried out investigations to check the validity of the equations suggested by Abraham on the one hand and by Lorentz and Einstein on the other.

First of all it is important to note that the mass of the *slow* electron is deduced from two sets of experimental data: (1) those involving a determination of e, the electric charge carried by the electron, and (2) those involving e/m_0, the specific electronic charge.

As a result of a critical survey of the considerable number of measurements which have been made of each of these constants, R. T. Birge[5] concludes that the most accurate values are as follows:*

e = $(1.60203 \pm 0.00034) \times 10^{-20}$ abs e.m.u.
e/m_0 = $(1.7592 \pm 0.0005) \times 10^7$ abs e.m.u.g.$^{-1}$
Hence m_0 = $(9.1066 \pm 0.0032) \times 10^{-28}$ g.

(5)Reports on Progress in Physics, 8, 90 (1941); Rev. Modern Phys., 13, 233 (1941).
*1 abs. e.m.u. = 1 absolute electromagnetic unit = 10 absolute coulombs.

Reprinted from G.E. Review (Oct. 1944 - Pages 6-13) By Permission

In the experimental investigations carried out to test the validity of Equation (4) it has been customary, therefore, to measure e/m_0 as a function of $\beta = v/c$.[a]

W. Kaufman[6] was one of the first investigators to carry out such a series of measurements. He worked with beta particles emitted by radioactive atoms, and his results were apparently not in satisfactory agreement with the Lorentz-Einstein prediction.

A. H. Bucherer[7] also used beta particles for which $\beta(=v/c)$ varied from 0.317 to 0.687, and obtained agreement with the Lorentz-Einstein relation. His results were questioned, however, by A. Bestelmeyer.[8]

E. Hupka[9] made use of electrons emitted, by an ultraviolet source, from copper, which were accelerated by voltages ranging from 17,430 volts (for which, as later shown, $\beta = 0.255$) to 88,400 volts ($\beta = 0.52366$). The experimental results were in accord with the relativity relation.

In 1914, G. Newmann,[10] working with Bucherer, and using the same experimental arrangement as that used by the latter in his 1909 investigation, obtained agreement with the relativity relation for beta particles ranging in velocity from $\beta = 0.4$ to $\beta = 0.7$.

C. E. Guy and C. Lavanchy[11] worked with cathode rays having velocities up to 0.48 that of light and found that the values observed for m/m_0 were in better agreement with the Lorentz-Einstein relation than with that deduced by Abraham.

R. A. R. Tricker,[12] working with beta rays from a thin deposit of radioactive material, which had velocities as high as 0.8 that of light, obtained results in very satisfactory agreement with the relativity relation.

Finally, in the most recent investigation by C. T. Zahn and A. H. Spees,[13] who used beta rays for which $\beta = 0.75$, the results obtained confirmed the conclusions of the previous investigators.

"All these experiments," according to J. D. Stranathan, "leave no doubt that the mass of an electron varies with velocity," and that the relation between m/m_0 and v is that deduced by Lorentz and Einstein.

A further interesting confirmation of this relation has been obtained from another type of investigation by H. R. Robinson and his collaborators.[b] When very short wavelength x-rays impinge on the surface of a metal, electrons are ejected with energies of sufficient magnitude to require the application of the relativity theory.

According to this theory, the kinetic energy of a particle having a velocity β of that of light is given by the relation

$$K = m_0 c^2 \left[1/(\sqrt{1-\beta^2}) - 1 \right] \tag{5}$$

'For low values of $\beta = v/c$

$$\left(1 - \frac{v^2}{c^2} \right)^{-1/2} = 1 + \frac{1}{2} \frac{v^2}{c^3} + \frac{3}{8} \frac{v^4}{c^4}$$

and consequently, if we neglect terms involving $(v/c)^4$ and higher powers

$$K = \frac{1}{2} m_0 v^2 \tag{6}$$

which is the classical mechanics definition of kinetic energy.

In Robinson's experiments the energy of a quantum of x-radiation (usually designated a photon), of frequency ν, has the magnitude $h\nu$, where h, the quantum constant, has the value, according to Birge,

$$h = (6.624 \doteq 0.002) \times 10^{-27} \text{ erg sec}$$

This energy is used up in (1) liberating an electron from one of the innermost shells of the atom, which we shall denote by $h\nu_A$, and (2) imparting to the emitted electron a kinetic energy, K. Hence

$$h(\nu - \nu_A) = K \tag{7}$$

On the other hand ν_A has been determined from measurements of the characteristic absorption and emission spectra in the x-ray region. The values of K were obtained from measurements of the radius of curvature of the ejected electron in a magnetic field. The energies actually observed varied from 200 to 350 kilovolts, and were found to be in excellent agreement with the predictions from Equation (7) when K was calculated by means of Equation (5).

We shall now consider more fully the quantitative significance of the above equations. When electrons emitted from the cathode of a cathode-ray tube or a Coolidge X-ray tube are accelerated by a potential of magnitude V volts, the kinetic energy acquired by the electrons is given by the relation

$$K = eV \times 10^8/c \text{ erg}$$

where c = velocity of light
= $(2.99776 \doteq 0.00004) \times 10^{10}$ cm sec^{-1}

Hence $\quad K = 1.60203 \times 10^{-12} V \text{ erg} \tag{8}$

where V is expressed in absolute volts.

Substituting this value for K, in Equation (5) and also substituting for m_0 and c the values just given, it follows that

$$m/m_0 = 1 + 1.9576 \times 10^{-6} V \tag{9}$$

and $\quad 1 - \beta^2 = (1 + 1.9576 \times 10^{-6} V)^{-2} \tag{10}$

(a) For a description of the methods used for the determination of the specific electronic charge, see Ref. (I) and (II) at the end of this article.
(b) For a detailed account of these experiments see p. 525 of Ref. (I) and p. 250 of Ref. II at the end of this article.
(6) *Physikal. Zeits.*, 4, 55 (1902); *Ann. d. Physik*, 17, 487 (1906).
(7) *Ann. d. Physik*, 28, 513 (1909).
(8) *Ann. d. Physik*, 30, 166 (1909), ib. 32, 231 (1910); also Bucherer, *Ann. d. Physik*, 30, 974 (1909).
(9) *Ann. d. Physik*, 31, 169 (1909).
(10) *Ann. d. Physik*, 45, 529 (1914).
(11) *Comptes Rendus*, 161, 52 (1915).
(12) *Proc. Roy. Soc.*, A, 109, 384 (1935).
(13) *Phys. Rev.*, 53, 357, 365 (1938).

For values of $V < 10^4$, m/m_0 is practically independent of the voltage and equal to 1.02 or less. Also, for low values of β, as shown in Equation (6),

$$v^2 = 2 \times 1.6020 \times 10^{-12} \, V/m_0$$

Hence

$$v = 5.9317 \times 10^7 \, \sqrt{V} \text{ cm sec}^{-1} \tag{11}$$

and

$$v/c = 1.9787 \times 10^{-4} \, \sqrt{V} \tag{12}$$

For very large values of V ($>10^7$)

$$m/m_0 = 1.9576 \times 10^{-3} \, V \tag{13}$$

and

$$1 - \beta = \frac{1}{2} \, (1.9576 \times 10^{-6} \, V)^{-2}$$

$$= 1.3047 \times 10^{11} \, V^{-2} \tag{14}$$

These relations have been used to calculate the data given in Table II.

Fig. 1 shows a log-log plot of values of β versus V. Above $V = 10^6$ volts, it is much more reasonable to plot values of $\log (1-\beta)$ against $\log V$, as shown in **Fig. 2**. There is also given a plot of $\log (m/m_0)$ against $\log V$. It will be observed that up to about $V = 10^4$, Equation

Interchangeability of Mass and Energy

From Equations (4) and (6) it follows that the kinetic energy,

$$K = (m - m_0) \, c^2 \tag{15}$$

while, the *total energy*

$$E = mc^2 = m_0 \, c^2 / \sqrt{1 - \beta^2} \tag{16}$$

Equation (15) suggests that the increase in energy is somehow associated with the increase in mass, and that even the rest mass m_0 is due to an internal energy of amount $m_0 c^2$. This may then be designated the *rest energy* of the particle. Also, we may regard the total energy E as equivalent to an inertial mass of magnitude

$$m = E/c^2 \tag{17}$$

As pointed out in Richtmyer and Kennard's book,[c] "The foregoing relations suggest that inertial mass may be a property of *energy* rather than of matter as such, each erg of energy possessing, or having associated with it, $1/c^2$ gm of mass. The law of conservation of mass

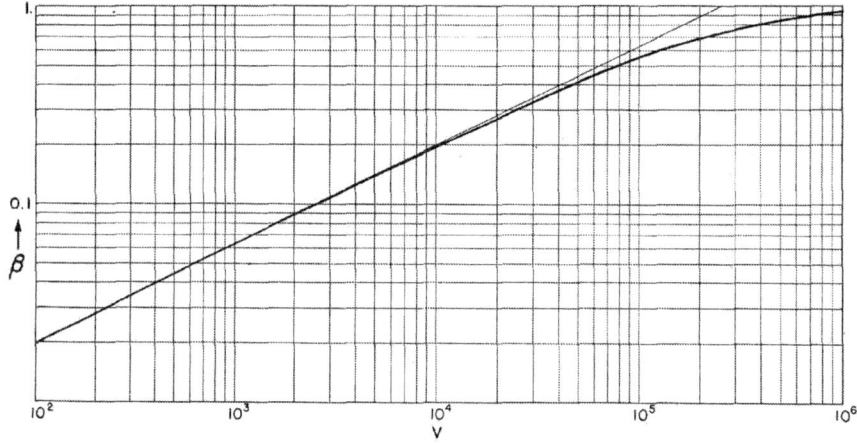

Fig. 1. Log-log plot of values of β versus V

(12) is valid, as shown by the linear portion of the plot in **Fig. 1**. Also, as shown in **Fig. 2**, the graphs for log $(1-\beta)$ and $\log (m/m_0)$ become practically linear for $V > 10^7$, which is in accord with Equations (13) and (14).

For example, in the two-million-volt x-ray tube, $\beta = 0.98$, that is, the velocity of the electrons is within 2 per cent of that of light, and $m/m_0 = 3.95$ approximately. At 100 million volts, which is obtained from the electron accelerator, $m/m_0 \approx 197$ approximately, and the velocity of the electrons is only 0.00129 per cent less than that of light.

would then become merely another aspect of the law of conservation energy."

This conclusion that mass and energy are interchangeable has been signally confirmed by the investigations in the field of nuclear reactions, as will be mentioned at greater length in a subsequent section. Before discussing the results obtained in these investigations, however, it is necessary to establish certain quantitative relations which follow from Equation (17).

[c]See p. 145 of Ref. (1) at the end of this article.

Since $c = 2.99776 \times 10^{10}$ cm sec^{-1}

$E = 8.9866 \times 10^{20}$ erg per gm

$\quad = 2.4963 \times 10^7$ kilowatt-hour per gm

$\quad = 8.5197 \times 10^{10}$ B.t.u. per gm

$\quad = 2.1468 \times 10^{13}$ gm-calories per gm of mass.

Thus, the energy equivalent to the annihilation of 1 gm mass is approximately 25 million kwh. An idea of the magnitude of this amount of energy can be obtained from the following figures: the Grand Coulee Development in Washington will ultimately supply 1.89 million kw, and that at Boulder Dam, 1.322 million kw. Thus, the energy equivalent of 1 gm mass represents the total energy from the larger power-source for a period of about 13 hours, and from the second one, for a period of about 19 hours.

It is of interest to compare the mass energy equivalent with that evolved in a chemical reaction. For instance, in the combination of two hydrogen atoms to form molecular hydrogen, according to the reaction,

$$H + H \longrightarrow H_2$$

about 100,000 gm-calories are evolved,[f] or about 50,000 calories per gm of hydrogen. The energy associated with 1 gm, on the basis of the relativity principle, is 4.294×10^8 times that evolved in the chemical reaction. Similarly in the combustion of carbon to form CO_2, the rest energy per gm of carbon is 2.72×10^9 times that evolved in the complete oxidation of the same amount of carbon. Thus, the inertial mass of a body is of the order of a *billion times* that evolved in any chemical reaction.

That no change in mass has been observed in chemical reaction is evidently due to the fact that the decrease in mass equivalent to the energy evolved in a

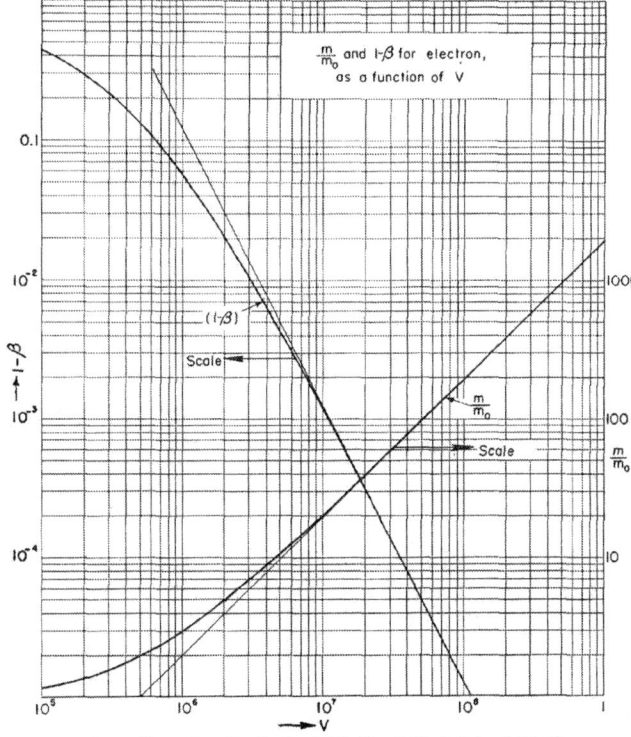

Fig. 2. Plot of values of log $(1 - \beta)$ against log V; and of log (m/m_0) against log V

chemical reaction is about 10^{-9} gm per gm of material—a quantity that is considerably less than can be measured by even the most sensitive type of balance.

However, in the case of reactions between atomic nuclei and very high velocity charged particles such as protons, deuterons, and alpha particles, and in the reaction of nuclei with gamma rays, the changes in energy are so large that changes in mass actually occur.

As an illustration let us consider the nature of the reaction which occurs when the nuclei of the isotope of lithium of mass 7 (designated by Li⁷) are bombarded by protons (nuclei of the hydrogen isotope of mass 1, that is, H¹). Two alpha particles (nuclei of the helium atom

(f) Atomic hydrogen welding depends upon the utilization of this energy.

TABLE II
MASS AND VELOCITY OF ELECTRON AS A FUNCTION OF ACCELERATING VOLTAGE

Volts	m/m_0	β	$1 - \beta$	v (cm sec^{-1})
10^2	1.0002	0.0198	0.9802	5.932×10^8
10^3	1.00196	0.0626	0.9374	1.876×10^9
10^4	1.0196	0.1950	0.8050	5.932
3×10^4	1.0587	0.3284	0.6716	9.845
5×10^4	1.0979	0.4128	0.5872	1.237×10^{10}
10^5	1.1958	0.5483	0.4517	1.644
3×10^5	1.5873	0.7766	0.2234	2.329
10^6	2.9576	0.9411	0.0589	2.821
3×10^6	6.8728	0.9894	0.0107	2.966
10^7	20.576	0.9982	1.181×10^{-3}	
10^8	196.76	approx 1.	1.292×10^{-5}	
10^9	1958.6	approx 1.	1.303×10^{-7}	

MASS-ENERGY RELATION

(Continued from preceding page)

of mass 4) are formed, and there is a *decrease* in mass which can be calculated from the atomic mass units of the corresponding elements, which are as follows:[†]

Li^7	7.01818 a.m.u.
H^1	1.00813
He^4	4.00389

The decrease in atomic mass units is 0.01853, which corresponds to 1.665×10^{19} ergs. This energy appears as kinetic energy of the alpha particles formed in the reaction.

It is customary in connection with nuclear reactions to indicate the energy equivalent of the mass in terms of electron volts. The relation between the two magnitudes is deduced as follows:

Since there are 6.0227×10^{23} atoms per gm-atomic mass of any element, a decrease in atomic mass of 1 a.m.u. corresponds to an energy gain per individual atom of

$$10^{-23} c^2/6.0227 = 1.661 \times 10^{-24} \times (2.99776 \times 10^{10})^2 \text{ erg}$$
$$= 1.493 \times 10^{-3} \text{ erg}$$
$$= 1.493 \times 10^{-3}/1.602 \times 10^{-12} \text{ electron volts}$$
$$= 931.5 \text{ Mev}$$
$$\text{or} \quad 1 \text{ Mev} = 1.0736 \times 10^{-3} \text{ a.m.u.} \quad (18)$$

Thus, the energy evolved in the reaction between Li^7 and H^1 to form two alpha particles corresponds to $0.01853 \times 931.5 = 17.26$ Mev.

Actually the observed *kinetic energy* of the alpha particles, in this reaction, is 8.64 Mev per particle, or 17.28 Mev for the reaction.

The verification by experimental observation of this equivalence of mass and energy is undoubtedly "one of the greatest triumphs of the theory of relativity."[d]

The confirmation of the Einstein relation would have been impossible if it had not been for the accurate measurement of nuclear masses by means of the mass spectrometer first developed by J. J. Thomson and F. W. Aston. In recent years, the latter, K. T. Bainbridge, A. J. Dempster, W. Bleakney, A. O. Nier, and others have carried out extremely exact measurements. As a result, the atomic masses of practically most of the lighter isotopes (from H^1 to Ca^{40}) are now known to a very high degree of accuracy.[e]

Table III gives the atomic masses of the isotopes which are mentioned in Table IV. The latter[f] gives a number of further illustrations of the validity of the energy-mass equivalence relation, as observed in nuclear disintegrations.

In Table III the designations of the four lightest nuclei are indicated. It should be observed that the neutron, discovered by J. Chadwick in 1932, is a

particle of practically the same mass as the proton but possessing zero charge. It may be regarded as constituted of an electron and proton.

Table IV also shows illustrations of nuclear reactions in which there is an increase in mass which is obtained from the decrease in kinetic energy of the incident alpha particle.

The last reaction mentioned in Table IV shows that the energy of a light quantum of frequency ν can also, under certain conditions, be interchanged with mass. In the reaction mentioned, incident gamma radiation breaks up the nucleus of "heavy" hydrogen (the deuteron) into a proton and a neutron with resulting increase in mass. From the value given for h in this article, it follows that the *frequency corresponding to 1 Mev energy* is

$$\nu = (1.602 \times 10^{-6})/(6.624 \times 10^{-27})$$
$$= 2.417 \times 10^{20} \text{ sec}^{-1} \quad (19)$$
$$\text{and} \quad \lambda = c/\nu$$
$$= 1.240 \times 10^{-10} \text{ cm}$$
$$= 1.240 \times 10^{-2} \text{ Å} \quad (20)^*$$

*1 Å = 1 Angstrom unit = 10^{-8} cm

Since 1 Mev is associated with a decrease in a.m.u. of 1.0736×10^{-3}, a decrease of 1×10^{-3} a.m.u. corresponds to radiation of wavelength, $\lambda = 1.240 \times 1.0736 \times 10^{-2} = 1.331 \times 10^{-2}$ Å.

Creation and Annihilation of Electron-Positron Pairs[z]

In 1932, C. D. Anderson announced the discovery of[14] the occurrence in cosmic rays of a charged particle of approximately the same mass as the electron but possessing a positive charge of the same magnitude as that of the electron. This particle has been designated the *positron*.

TABLE III

ATOMIC WEIGHTS OF SELECTED ISOTOPES

Atom	Atomic Weight	Atom	Atomic Weight
n^1 (neutron)	1.00893	B^{10}	10.01631
H^1 (proton)	1.00813	C^{11}	11.01526
H^2 (deuteron)	2.01473	C^{12}	12.00398
He^4 (alpha particle)	4.00389	N^{14}	14.00750
Li^6	6.01686	O^{17}	17.00450
Li^7	7.01818	Si^{28}	27.98639
Be^9	9.01504	P^{31}	30.98457

TABLE IV

COMPARISON OF CALCULATED AND OBSERVED ENERGIES OF DISINTEGRATION

	Decrease of Mass	ENERGY RELEASED, MEV	
		Calculated	Observed
Be^9 + H^1 ⟶ Li^6 + He^4	0.00242	2.25	2.28
Li^6 + H^2 ⟶ He^4 + He^4	0.02381	22.17	22.20
B^{10} + H^2 ⟶ C^{11} + n^1	0.00685	6.38	6.08
N^{14} + H^2 ⟶ C^{12} + He^4	0.01436	13.37	13.40
N^{14} + He^4 ⟶ O^{17} + H^1	−.00124	−1.15	−1.16
Si^{28} + He^4 ⟶ P^{31} + H^1	−.00242	−2.25	−2.23
H^2 + hν ⟶ H^1 + n^1	−.00233	−2.17	−2.17

[†] These atomic masses include the masses of the extra-nuclear electrons which, of course, remain constant during the reaction.
[d] See p. 72 of Ref. (III) at the end of this article.
[e] For a complete list of these isotopes and their atomic masses, see p. 447 of Ref. (II) and p. 231 of Ref. (III) at the end of this article.
[f] These were selected from a much larger number of reactions for which Stranathan gives a comparison of calculated and observed energies of disintegration. See pp. 446–447 of Ref. (II) at the end of this article.

[z] This section is based on the discussion of this topic in Ref. (II) at the end of this article.
[14] *Science*, 76, 238 (1932); *Phys. Rev.*, 43, 491 (1933).

MASS-ENERGY RELATION
(Continued from preceding page)

There is considerable evidence that an electron and a positron can combine with the result that the two masses disappear and there appears simultaneously a radiation of a frequency, ν, and wavelength, λ, which are in agreement with the predictions of the last two equations. There is also evidence that a quantum of gamma radiation of the proper frequency ν can disappear and in its place an electron-positron pair will be formed.

The rest energy (in Mev) corresponding to the mass, m_0, of an electron is

$$(9.1066 \times 10^{-28})(6.0227 \times 10^{23})/(1.0736 \times 10^{-3}) = 0.511 \text{ Mev}$$

Hence, for an electron-positron pair, the rest energy is equivalent to 1.022 Mev, and in accordance with Equation (20), it would be expected that if in the disappearance of a pair, a single quantum of radiation appears, the wavelength of the radiation should be $1.240 \times 10^{-2}/1.022 = 1.211 \times 10^{-2}$ Å. If, on the other hand, the energy is split between two quanta, each would correspond to radiation of wavelength 0.024 Å.

As a matter of fact, "Several experimenters have found radiation of these two wavelengths formed under conditions where there should be a number of free positrons combining with electrons." Furthermore, "this annihilation radiation has also been observed coming from artificial radioactive bodies which emit positrons, and also from metals against which the positrons are allowed to strike."[h]

The inverse process has also been investigated. Gamma rays from ThC "eject positrons as well as electrons when they fall upon heavy elements such as lead. These radiations have an energy of 2.62 Mev, while the sum of the energies of the emitted positrons and electrons has a maximum value of 1.6 Mev. Obviously the difference, that is, 1.02 Mev, is that required to form the rest masses of the particles."

Origin of Energy in Stars

The problem of the origin of the energy emitted from the sun and other stars has confronted physicists for well nigh over a century and a half. Various hypotheses have been suggested, but none has turned out completely satisfactory. However, in 1939, H. A. Bethe published a paper[18] in which an apparently satisfactory explanation is given in terms of nuclear reactions and of the law of equivalence of mass and energy.

As shown by Bethe, the most important source of energy in ordinary stars is the reaction of carbon and nitrogen with protons according to the following cycle:

$$C^{12} + H = N^{13} + \text{Gamma radiation } (\gamma)$$
$$N^{13} + \gamma = C^{13} + \text{Positron}$$
$$C^{13} + H = N^{14} + \gamma$$
$$N^{14} + H = O^{15} + \gamma$$
$$O^{15} = N^{15} + \text{Positron}$$
$$N^{15} + H = C^{12} + He^4$$

(h) See pp. 375, 377 of Ref. (II) at the end of this article.
(18) Phys. Rev., 55, 434 (1939).

The net result of the cycle is the *formation of an alpha particle from four protons*, in accordance with the reaction

$$4H = He^4 + 2 \text{ Positrons} + \gamma\text{-radiation}$$

The isotope of carbon C^{12} acts as a "catalyst," since it is reproduced at the end, and the actual reaction consists in the *combination of four protons and two electrons to form an alpha particle*.[†]

From Table III giving the masses of the isotopes, it follows that the decrease in mass which occurs in this reaction is equivalent to approximately 1×10^{-15} erg per individual proton.

At the temperature in the center of the sun, which is approximately 20,000,000 C, the foregoing reaction is the only one which, according to Bethe's argument, is capable of yielding the observed energy emission, 2 ergs per gm per sec. The sun's atmosphere contains 35 per cent hydrogen, and on the basis of the energy evolution observed, Bethe calculates that "the prospective future life of the sun should be 12.10^9 years."

Bethe has also shown that in the case of fainter stars, with lower central temperature, the reaction

$$H + H = H^2 + \text{Positron}$$

is probably the main source of energy

Conclusion

In this article abundant experimental evidence has been given for the two relations which follow from the principle of relativity. The first of these states that the mass of a particle increases with velocity until it becomes infinitely great for speeds that approach that of light.

The second relation states that energy and mass are equivalent. As has been pointed out, the agreement between calculated and observed energies of nuclear transformation is ample confirmation of this conclusion. As Stranathan remarks, "One would be scarcely more justified today in questioning the interchangeability of mass and energy than he would be in questioning the interchangeability of heat and energy."

What all this signifies for the future is a problem of greatest importance to the future of the human race itself. Will man utilize this new knowledge for peace and progress beyond the dreams of even the most daring optimist? Or will this be the harbinger of a Frankenstein that will destroy its own creator? Will the human race profit by past experience?

(†) The nucleus of He has a charge of two positive units. Hence the four protons, each of which has a nuclear charge of one positive unit, must give up two positive units in the reaction, or combine with two electrons.

Additional References

(I) F. K. Richtmyer and E. H. Kennard, *Introduction to Modern Physics*, McGraw-Hill Book Company, Inc., 1942.

(II) J. D. Stranathan, *The "Particles" of Modern Physics*, The Blakiston Company, Philadelphia, 1942.

(III) E. Pollard and W. L. Davidson, *Applied Nuclear Physics*, John Wiley and Sons, Inc., New York, 1942.

NEW EXPERIMENTAL TEST OF SPECIAL RELATIVITY

J. P. Cedarholm, G. F. Bland,
and B. L. Havens
International Business Machines
Watson Laboratory at Columbia University,
New York, New York

and

C. H. Townes
Department of Physics,
Columbia University, New York, New York
(Received September 29, 1958)

The relative frequency stability of two beam-type maser oscillators is used to test the dependence of the velocity of light on velocity of the frame of reference with considerably more precision than has been obtained from experiments of the Michelson-Morley[1] type. Expressed in terms of an ether, the maximum ether drift is shown to be less than 1/1000 of the earth's orbital velocity.

The experiment, which was performed at the Watson Laboratory, involves comparison of the frequencies of two masers[2] having their beams of NH_3 molecules traveling in opposite directions. Møller[3] has analyzed this case and given the change in frequency of a beam-type maser due to ether drift, assuming the molecules in the beam to have a velocity u with respect to the cavity through which they pass, and the cavity to have a velocity v with respect to the ether. The shift may be simply discussed by assuming that, if v is zero, radiation is emitted perpendicularly to the molecular velocity so that there is no Doppler shift. If the cavity and beam are then transported at velocity v through the ether in a direction parallel to u, radiation must be emitted by the molecules slightly forward at an angle $\theta = \pi/2$ $-v/c$ with respect to u. The fractional change in frequency due to the Doppler effect is then $\epsilon = u/c \cos\theta$ or uv/c^2 due to motion through the ether, assuming that the proper molecular frequencies are unchanged by such motion.

For a thermal molecular velocity of 0.6 km/sec and for the earth's orbital velocity (30 km/sec), $\epsilon = 2 \times 10^{-10}$. The difference in frequency due to the above effect between two masers with oppositely directed beams would be $2\epsilon\nu$, or about 10 cps for ν equal to 23 870 Mc/sec, the NH_3 inversion frequency.

Although uv/c^2 is of second order in the velocities, it is of first order in the velocity of the cavity, or of the laboratory, with respect to the

ether. The present experiment measures the entire effect with a rather small fractional error, which affords a particularly small upper limit to v since this quantity enters in first order, rather than in second order as in the Michelson-Morley experiment. A somewhat similar term would occur in the latter experiment if the interferometer used were transported by a plane of speed u, and interference fringes were compared for two opposite directions of flight.

Two maser oscillators with oppositely directed beams were mounted with necessary auxiliary equipment on a rack which could be rotated about a vertical axis. The beat frequency between the two oscillators was adjusted to about 20 cps and recorded continuously. After approximately one minute of recording with the maser axes oriented in an east-west direction, the apparatus was rotated 180° and the beat frequency recorded in the new position.

The change in beat frequency, on the basis of an ether drift, should be $4\epsilon\nu$, or about 20 cps. Sixteen such comparisons were made during a period of about 20 minutes. These were repeated about once per hour during a time somewhat longer than 12 hours, so that the earth's rotation would sweep the east-west direction through a plane.

A relative change in frequency of the two oscillators amounting to about 1 cps was found when they were rotated through 180°. This change is largely due to the earth's magnetic field and other local magnetic fields from which no shielding was attempted. The significant observation is that this change was independent of the time of day (or orientation of the earth), as indicated in Fig. 1.

The first series of measurements was made during a week-day, when local magnetic fields and line voltages were varying. It showed some systematic variations in the effect measured as large as ±1/20 cps during the day. A second series of measurements, taken on a Saturday when local disturbances were less serious, showed no variation greater than ±1/50 cps as indicated in Fig. 1, and even these appear random and not simply correlated with time (or the earth's orientation). This precision corresponds to a comparison of frequencies of the two masers to one part in 10^{12}.

The results show that any term of the form uv/c^2 must be smaller by a factor of at least 1000 than what would be predicted by setting v equal to the earth's orbital velocity. That is, velocity with respect to an ether in a plane per-

Reprinted from Phys. Rev. Letters
(Nov. 1958 - Pages 342-343)
By Permission

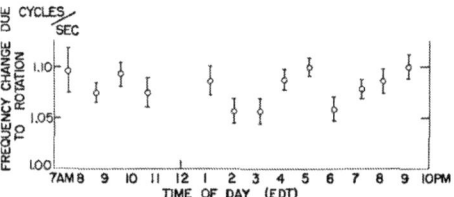

FIG. 1. Diurnal variation of the change in relative
frequency due to rotating two ammonia masers through
180°

Beams of the two masers were oppositely directed
and in an east-west direction. The change of about
1.08 cps is primarily due to local magnetic fields.
Maximum deviation from this value during the day is
1/50 cps. Lengths of lines indicate probable errors
computed from fluctuations of 16 measurements at each
point.

pendicular to the earth's axis must be less than
1/30 km/sec. Results from experiments of the
Michelson-Morley type vary from an ether drift
of about 8 km/sec reported by Miller[4] to an upper
limit of 1.5 km/sec given by the experiments of
Joos.[5] Of course a major part of the advantage of
the present experiment is its first-order rather
than second-order dependence on v.

Those who are already completely convinced of
the correctness of special relativity, or who do
not wish to consider an ether model, should note
that postulates of special relativity are not necessarily inconsistent with the existence of a frequency shift in the above experiment or of an
anisotropy in space. These can result from the
presence of matter external to the earth which
is not uniformly distributed, or which is not
moving with the earth's velocity.

The preliminary results quoted here apply to
September 20, 1958. It is expected that the experiment will be refined further and that additional measurements will be made at other times
during the year.

[1] A. A. Michelson and E. W. Morley, Am. J. Sci.
34, 333 (1887).
[2] Gordon, Zeiger, and Townes, Phys. Rev. 99, 1264
(1955).
[3] C. Møller, Suppl. Nuovo cimento 6, 381 (1957).
[4] D. C. Miller, Revs. Modern Phys. 5, 203 (1933).
See also Shankland, McCuskey, Leone, and Kuerti,
Revs. Modern Phys. 27, 167 (1955).
[5] G. Joos, Ann. Physik 7, 385 (1930).

Variation of the Rate of Decay of Mesotrons with Momentum

Bruno Rossi* and David B. Hall
University of Chicago, Chicago, Illinois
(Received December 13, 1940)

In order to determine the dependence of the probability of decay on momentum, mesotrons with range between 196 and 311 g/cm² of lead and mesotrons with range larger than 311 g/cm² of lead were investigated separately. The softer group of mesotrons was found to disintegrate at a rate about three times faster than the more penetrating group, in agreement with the theoretical predictions based on the relativity change in rate of a moving clock. A new value of the proper lifetime of mesotrons of $(2.4\pm0.3)\times10^{-6}$ sec. is determined, based upon measurements with particles with momentum of approximately 5×10^8 ev/c.

Introduction

RECENT experiments on the variation of cosmic-ray intensity with altitude have shown that the rate of decrease of the mesotron component with increasing atmospheric depth cannot be accounted for completely by ordinary ionization losses. It has been established, namely, that the number of mesotrons is much more strongly reduced by a layer of air than by a layer of condensed material which is equivalent to the air layer with regard to ionization losses.[1-5]

The anomalous absorption in air is interpreted on the hypothesis that mesotrons disintegrate spontaneously with a proper lifetime of the order of a few microseconds. According to this assumption, a considerable fraction of the mesotron beam will disappear by disintegration while traveling in the atmosphere. No appreciable number of mesotrons, however, will disintegrate within a condensed absorber, even equivalent in mass to the whole thickness of the atmosphere, because the time required for the traversal of such an absorber is very short compared with the lifetime of mesotrons.

A simple relativistic consideration shows that if the absorption anomaly of mesotrons is due to spontaneous decay it must be more pronounced for mesotrons of low energy than for mesotrons of high energy. In fact, let τ_0 be the "proper lifetime" of mesotrons; i.e., the lifetime measured in a frame of reference in which the mesotron is at rest, and τ the lifetime measured in a frame of reference in which the mesotron is moving with a velocity β.[6] Then

$$\tau=\tau_0/(1-\beta^2)^{\frac12} \qquad (1)$$

and the "average range before decay" L; i.e., the average distance traveled by the mesotrons before disintegrating, becomes

$$L=\beta\tau=p\tau_0/\mu, \qquad (2)$$

* Now at Cornell University, Ithaca, New York.
[1] B. Rossi, N. Hilberry and J. B. Hoag, (a) Phys. Rev. 56, 837 (1939); and (b) Phys. Rev. 57, 461 (1940).
[2] W. M. Nielsen, C. M. Ryerson, L. W. Nordheim and K. Z. Morgan, Phys. Rev. 57, 158 (1940).
[3] M. Ageno, G. Bernardini, N. B. Cacciapuoti, B. Ferretti and G. C. Wick, Phys. Rev. 57, 945 (1940).
[4] H. V. Neher and H. G. Stever, Phys. Rev. 58, 766 (1940).
[5] A. Ehmert, Zeits. f. Physik 115, 333 (1940).
[6] We shall use throughout the paper the system of units described by B. Rossi, Phys. Rev. 57, 660 (1940).

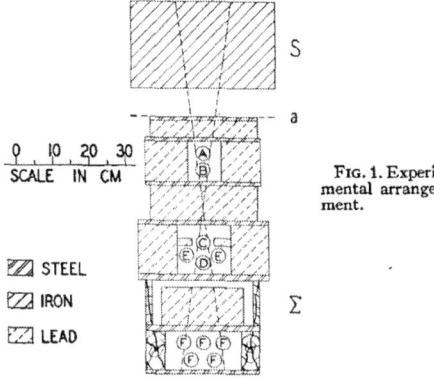

SCALE IN CM
0 10 20 30

STEEL
IRON
LEAD

FIG. 1. Experimental arrangement.

where μ is the mass and $p=\mu\beta/(1-\beta^2)^{\frac{1}{2}}$ is the momentum of the mesotrons. The probability of decay per centimeter path is obviously $1/L$. It is seen that the average range is directly proportional, and the probability of decay inversely proportional, to the momentum.

The experiments described in the present paper were primarily designed to test the dependence of disintegration probability on momentum expressed by Eq. (2). The purpose was to provide an additional check of the disintegration hypothesis and simultaneously to verify the relativistic transformation formula for time intervals. Further experimental evidence on the subject of the decay was particularly desirable in view of Fermi's recent theory showing that the energy losses of fast particles in condensed materials are appreciably reduced by the dielectric polarization of the medium.[7] According to this theory even *stable* mesotrons are absorbed by gases more strongly than by solid or liquid materials of the same mass per cm². The difference in absorption due to polarization increases with increasing mesotron momentum; i.e., varies oppositely from the difference due to decay. The polarization effect, as calculated by Fermi, was quantitatively inadequate to account for the experimental results already obtained. Yet it was interesting to investigate whether the observed absorption anomaly was a decreasing function of the mesotron momentum, as the anomaly attri-

[7] E. Fermi, Phys. Rev. **57**, 485 (1940).

buted to the decay, or an increasing function, as the anomaly attributed to the polarization.

An attempt to determine the rates of decay of mesotrons of different momenta has been reported by Nielsen, Ryerson, Nordheim and Morgan.[2] Mesotron groups of different average momentum were selected by taking the difference between the intensities of the mesotron beam after filtration through various thicknesses of lead. By this method, however, it is hardly possible to reach a sufficient accuracy, since the difference is small compared with the quantities directly measured. In our experiments the statistical precision was greatly improved by measuring the difference directly; i.e., by recording only mesotrons which can traverse a given thickness of lead but are stopped by a certain additional absorber.

EXPERIMENTAL METHOD

The experimental arrangement is schematically represented in Fig. 1. The Geiger-Müller counter tubes were of the self-quenching type. Their internal diameter was 4 cm and their effective lengths were as follows: counters A, B, C and E, 27 cm; counter D, 20 cm; counters F, 60 cm. The five counters F and the two counters E were all connected in parallel. The counter battery F covered completely the solid angle subtended by counters A, B, C and D. In order to cut off the soft component, 5 cm of lead was permanently placed above counter A and 10 cm of lead between counters B and C. Including the material of the frame, the permanent absorber above or between counters A, B, C and D was equivalent in absorption power to 186 g/cm² of lead, while that between D and F was equivalent to 10 g/cm² of lead. Counters A, B, C and D were protected on the side by lead walls 11.5 cm thick. An additional lead absorber Σ of 115 g/cm² could be introduced between D and F and an absorber S made of iron plates could be arranged above the apparatus so as to cover the whole solid angle subtended by counters A, B, C and D. The apparatus was set up in a moving van which could be taken to different altitudes on mountain roads. The whole system, except for the absorber S, was enclosed in a thermostatic box.

By means of an appropriate vacuum-tube

circuit, the following events were simultaneously recorded: (1) Fivefold coincidences between counters A, B, C, D and one of counters F or E (coincidences $[ABCD(E+F)]$); (2) coincidences between counters A, B, C and D not accompanied by a pulse either of counters F or of counters E (anticoincidences $[ABCD-(E+F)]$). The coincidences $[ABCD(E+F)]$ were mainly caused by mesotrons going through counters A, B, C, D and F. After entering the apparatus; i.e., after crossing the surface indicated by a in Fig. 1, these mesotrons had to traverse 196 g/cm² of lead when there was no absorber in Σ, or 311 g/cm² of lead when 115 g/cm² of lead were placed in Σ. Chance coincidences were negligible and coincidences due to air showers were certainly rare on account of the heavy lead shield at the side of the counters. Coincidences caused by ionization showers generated by mesotrons in the various absorbers could not introduce any error because they were a small and *constant* fraction of the coincidences caused directly by mesotron traversals.[8] Thus, one is justified in taking the counting rate $[ABCD(E+F)]$ as a measure of the number N of mesotrons entering the apparatus with a residual range larger than the total amount of matter present above or between the counters.

Anticoincidences $[ABCD-(E+F)]$ could be accounted for by one of the following events. (a) A mesotron has traversed A, B, C and D and has been stopped between D and F. (b) A mesotron has gone through A, B, C, D and F, but the counter battery F has failed to detect it for lack of efficiency. (c) A chance coincidence between pulses of counters A, B, C and D has occurred. (d) A mesotron has traversed counters A, B, C and D, but has been scattered through a wide angle so as to miss the counter battery F.

The stopping of mesotrons between D and F (case (a)) is certainly the main origin of the anticoincidences recorded with lead in Σ, which are, as we shall see, several times more frequent than those recorded without lead. The events described under (b) and (c) are about equally frequent with and without the absorber Σ. The scattering (case (d)) may contribute a small

number of anticoincidences, which is not necessarily the same with and without the absorber. Since, however, only slow mesotrons are appreciably scattered as well as absorbed, the difference in the number of anticoincidences due to scattering is a small and constant fraction of the difference in the number of anticoincidences due to absorption. Thus, the difference between the number of anticoincidences recorded with and without lead in Σ is proportional, if not accurately equal, to the number of mesotrons which traverse 10 g/cm² of lead and are absorbed by 125 g/cm² of lead between D and F. These mesotrons are those which enter the apparatus with a residual range between $R_a = 196$ and $R_b = 311$ g/cm² of lead.

THE MEASUREMENTS

Measurements were taken alternately at Denver, Colorado, and at Echo Lake, approximately 30 miles west of Denver. The geomagnetic latitude is practically the same (49° N) for both locations. The difference in altitude is 1624 m. The difference in atmospheric pressure, as measured during the experiments, was 108 mm Hg, equivalent to 147 g/cm².

The measurements at Echo Lake were performed partly with an iron absorber of 200 g/cm² in S and partly without this absorber. No iron absorber was used at Denver. Three complete sets of measurements were carried out at Denver and two at Echo Lake. The deviations of the single readings from the averages were within the statistical fluctuations. The final results are summarized in Table I. The errors given are the standard statistical deviations.

TABLE I. *Summary of the measurements at Denver and Echo Lake.* $[ABCD(E+F)]$ *and* $[ABCD-(E+F)]$ *are the numbers of coincidences and anticoincidences per minute.* Δ *is the difference between the numbers of anticoincidences per minute recorded with and without 115 g/cm² of lead in* Σ. *The errors are the standard statistical deviations.*

LOCATION	ABSORBER g/cm² S(Fe)	ABSORBER g/cm² Σ(Pb)	TIME MIN.	[ABCD ×(E+F)]	[ABCD −(E+F)]	Δ
Denver z=1616 m h=856 g/cm²	0	0	3384	5.16±0.048	0.091±0.0052	0.275
	0	115	6783	4.79±0.027	0.367±0.0074	±0.009
Echo Lake z=3240 m h=709 g/cm²	0	0	308	6.87±0.15	0.15 ±0.02	0.53
	0	115	1469	0.49±0.066	0.68 ±0.021	±0.03
	200	0	2846	5.73±0.045	0.119±0.0064	0.394
	200	115	5362	5.43±0.032	0.513±0.0098	±0.012

---

[8] This was not always the case for the experimental arrangements previously used. See the discussion on p. 464, reference 1(b).

According to the discussion in the foregoing section, the counting rates $[ABCD(E+F)]$ with 115 g/cm² of lead in Σ can be taken as a measure of the number N of mesotrons entering the apparatus with a residual range larger than $R_b = 311$ g/cm² of lead, while the figures listed under Δ can be taken as a measure of the number n of mesotrons entering the apparatus with a residual range between $R_a = 196$ and $R_b = 311$ g/cm² of lead.

Let N_1, N_1' and N_2 be the values of N at Echo Lake under 200 g/cm² of iron, at Echo Lake without the iron absorber and at Denver without the iron absorber, respectively. Let n_1, n_1' and n_2 be the corresponding values of n. Considering first the measurements taken without the iron absorber, we have

$$n_1'/N_1' = 0.082 \pm 0.005, \quad n_2/N_2 = 0.058 \pm 0.002.$$

It appears that the fractional number of slow mesotrons increases rapidly with altitude, in agreement with the results of the absorption measurements in carbon by Rossi, Hilberry and Hoag.[1] Because of a possible effect of scattering on the determination of n, the above figures cannot be trusted to represent accurately the absolute values of the ratios n_1'/N_1' and n_2/N_2. However, the ratios between values of $[ABCD(E+F)]$ or Δ at different depths should not be appreciably affected by scattering or by other disturbing effects. Thus we have

$$N_2/N_1 = 0.883 \pm 0.007 \quad n_2/n_1 = 0.698 \pm 0.031$$
$$N_2/N_1' = 0.738 \pm 0.009 \quad n_2/n_1' = 0.520 \pm 0.035$$

where the actual errors should not exceed the statistical errors indicated.

DISCUSSION

In order to discuss our experimental results, we need a relation between ranges and momenta for mesotrons. The *momentum* loss due to collision is given by the Bethe-Bloch formula

$$-\frac{dp}{dx} = 2\pi r_0^2 NZ\mu_e \frac{1}{\beta^3}\left[\log\frac{W_m\mu_e\beta^2}{I^2Z^2(1-\beta^2)} + 1 - \beta^2\right], \quad (3)$$

where r_0 is the classical radius of the electron, N the number of atoms per cm³, Z the atomic number, μ_e the rest energy of the electron, β the velocity of the mesotron, W_m the maximum

transferable energy, and $I = 13.5$ ev (this expression differs by a factor $1/\beta$ from the expression for the *energy* loss). A correction has to be applied to account for the polarization effect pointed out by Fermi. The correction, however, is very small for the mesotron momenta in which we are interested. According to some recent calculations of Halpern and Hall, it is of the order of 2 percent for iron and of 3 percent for lead.[9] Numerical integration of the equation for the momentum loss yields the range as a function of the momentum. The ranges $R_a = 196$ g/cm² of lead and $R_b = 311$ g/cm² of lead, which define the mesotron groups considered in the present experiments, are thus found to correspond to the momenta $p_a = 3.1 \times 10^8$ and $p_b = 4.5 \times 10^8$ ev/c, respectively. Mesotrons reaching Denver with momenta equal to p_a and p_b had momenta equal to 5.9×10^8 and 7.3×10^8, respectively, at the altitude of Echo Lake. For mesotrons with momentum between 3.1×10^8 and 7.3×10^8 ev/c the ratio between momentum losses per g/cm² of air and of iron is very nearly a constant and equal to 1.23. Thus, as far as collision losses are concerned, 200 g/cm² of iron is approximately equivalent to 147 g/cm² of air.[10] Consequently, if the mesotrons were stable, one should observe the same mesotron intensity at Echo Lake under 709 g/cm² of air plus 200 g/cm² of iron as at Denver under 856 g/cm² of air alone. This applies to the mesotron band between 3.1×10^8 and 4.5×10^8 ev/c as well as to the whole mesotron spectrum above 4.5×10^8 ev/c.

Our experimental results show that both N and n are larger at Echo Lake under the iron absorber than at Denver without this absorber. The difference is accounted for by the decay of mesotrons on their way down from 3240 m to 1616 m. Let us define the *probability of survival* w_{12} between two elevations z_1 and z_2 as the

[9] See O. Halpern and H. Hall, Phys. Rev. 57, 459 (1940). We are greatly indebted to the authors for kindly communicating to us the numerical results of their calculations, which are not yet published.
[10] The thickness of the iron absorber is actually slightly larger than it should be. When the experiments were performed, the results of Halpern and Hall on the polarization effect were not yet available and previous calculations had given a larger correction for this effect (see reference (7)). However, both the coincidences $[ABCD(E+F)]$ and the anticoincidences $[ABCD-(E+F)]$ change so slowly with the thickness of the absorber S, that it is hardly necessary to apply any correction to the experimental results.

TABLE II. *Comparison between various determinations of the average range before decay L from measurements on the absorption anomaly for vertical mesotrons; z_1 is the elevation of the higher station, p_2 the momentum of the recorded mesotrons, p the effective momentum. The data of Rossi and Hall for $p_2 > 3.0 \times 10^8$ ev/c have been obtained from Table I, adding the counting rates [ABCD(E+F)] and [ABCD −(E+F)]. This sum represents the number of mesotrons with range > 186 g/cm² of lead.*

AUTHORS	z_1 METERS	p_2 10^8 EV/C	p 10^8 EV/C	L 10^5 CM	COMPENSATING ABSORBER
Rossi et al.[1]	4300	>2.5	>3.3	9.4±0.9	carbon above the counters
	3240	>2.5	>3.3	9.9±1.2	
	1616	>2.5	>3.3	9.5±1.7	
	180	>2.5	>2.5	9.4±1.6	
Nielsen et al.[2]	2040	>1.8	>3.5	6.2±0.5	carbon above the counters
	2040	>3.5	>5.3	8.9±1.2	
	2040	>5.9	>7.8	15.1±3.5	
	2040	1.8→3.5	3.5→5.3	2.45	
	2040	3.5→5.9	5.3→7.8	2.26	
Ageno et al.[3]	3460	>2.2	>4.5	31±5	lead between the counters
Rossi and Hall	3240	>3.0	>4.3	12.3±0.6	iron above the counters
	3240	>4.5	>5.8	13.3±0.9	
	3240	3.1→4.5	4.4→5.8	4.5±0.6	

probability that a mesotron present at the higher level z_1 does not disintegrate before reaching the lower level z_2. Then $w_{12} = n_2/n_1 = 0.698$ is the experimental value for the average probability of survival between $z_1 = 3240$ m and $z_2 = 1616$ m of the mesotrons which reach z_2 with momenta between p_a and p_b, and $W_{12} = N_2/N_1 = 0.883$ is the corresponding value for the mesotrons which reach z_2 with momenta larger than p_b. One sees that w_{12} is much smaller than W_{12}, which shows that slow mesotrons disintegrate at a much faster rate than the more energetic ones. This result is in agreement with the predictions based upon the disintegration hypothesis (see Eq. (2)) and affords strong support to the hypothesis itself. For a mono-energetic group of mesotrons, the probability of decay has a very simple theoretical expression, provided the momentum loss in the air layer between z_1 and z_2 can be neglected. In this case, Eq. (2) gives

$$\log w_{12} = -(z_1 - z_2)/L. \quad (4)$$

It is convenient to use Eq. (4) as a definition of L also when the momentum loss cannot be neglected. It can easily be proved that L is still related to the lifetime τ_0 by an expression of the type of Eq. (2)

$$L = p\tau_0/\mu, \quad (2)$$

where, however, p has now the following

meaning:

$$p = (p_2 + ah_2)$$
$$\times \left[1 + \log \frac{p_2 + a(h_2 - h_1)}{p_2} \middle/ \log \frac{h_2}{h_1} \right]^{-1}. \quad (5)$$

Here h_1 and h_2 are the atmospheric depths at the elevations z_1 and z_2, p_2 is the momentum of mesotrons at z_2 and a is the momentum loss per g/cm² of air.[11] The momentum p is intermediate between the initial momentum $p_1 = p_2 + a(h_2 - h_1)$ and the final momentum p_2. We shall refer to it as the *effective momentum.*

Let us first consider the experimental results concerning mesotrons which have residual momenta between 3.1×10^8 and 4.5×10^8 ev/c at the lower elevation. The corresponding effective momenta are 4.4×10^8 and 5.8×10^8 ev/c and we may take 5.0×10^8 ev/c as an average. For this mesotron group the experimental value of the probability of survival between 3240 and 1616 m is $w_{12} = 0.698 \pm 0.031$, and therefore $L = (4.5 \pm 0.6) \times 10^5$ cm. It then follows from (2): $\tau_0/\mu = (9.07 \pm 1.3) \times 10^{-4}$ cm c/ev and accordingly, taking $\mu = 8 \times 10^7$ ev/c², $\tau_0 = (7.2 \pm 0.9) \times 10^4$ cm/c, or $\tau_0 = (2.4 \pm 0.3) \times 10^{-6}$ sec. We shall next consider the continuous mesotron spectrum which reaches 1616 m with a residual momentum larger than 4.5×10^8 ev/c. The probability of survival for this mesotron group is $W_{12} = 0.883$, and if we take Eq. (4) as an experimental definition of L we get $L = (13.3 \pm 0.9) \times 10^5$ cm. Assuming $\tau_0 = 2.4 \times 10^{-6}$ sec., we then calculate formally, from Eq. (2), $p = 1.5 \times 10^9$ ev/c. This momentum should represent a sort of average effective momentum for the mesotron group considered. The value $p = 1.5 \times 10^9$ ev/c is quite compatible with our present knowledge of the momentum spectrum of mesotrons.[12] Thus, while a quantitative proof of Eq. (2) is still wanting, its approximative validity may be considered as established.

In evaluating the experimental results we have only considered mesotrons coming in the vertical direction. As a matter of fact, our experimental arrangement was strongly selective for vertical

[11] This follows immediately, for instance, from Eq. (14) of the paper "The disintegration of mesotrons" by B. Rossi, Rev. Mod. Phys. **11**, 296 (1939).
[12] Cf., e.g., P. M. S. Blackett, Proc. Roy. Soc. A**159**, 1 (1937); D. J. Hughes, Phys. Rev. **57**, 592 (1940).

mesotrons, but detected also mesotrons coming in directions inclined up to an angle of almost 45°. The inclined mesotrons travel a longer distance and have on that account a smaller probability of survival than the vertical ones. The increase in the path length, however, is partially compensated by an increase in the effective momentum. Thus the correction is not large and can be disregarded at the present state of the experimental accuracy.

COMPARISON WITH PREVIOUS RESULTS

Table II summarizes the data on the mesotron decay which can be deduced from the measurements so far reported on the absorption anomaly for vertical mesotrons. L is calculated, according to Eq. (4), from the experimental values of the probability of survival. The results of Nielsen, Ryerson, Nordheim and Morgan on the mesotron groups with p_2 from 1.8×10^8 to 3.5×10^8 ev/c and from 3.5×10^8 to 5.9×10^8 ev/c are not accurate enough for a quantitative comparison with our data on the mesotron group with p_2 from 3.1 ± 10^8 to 4.5×10^8 ev/c. No other measurements on selected groups of mesotrons are available. All the remaining data in Table II refer to mesotrons for which only the lower limit of the momentum is defined. A comparison between these data is not straightforward because L depends not only on the minimum effective momentum p_{min} of the mesotrons recorded, but also on the shape of the momentum spectrum, which is probably different at different altitudes. One may expect, however, an approximate correlation to exist between the values of p_{min} and L in the various experiments. Table II shows that this is actually the case if we exclude the measurements of Ageno, Bernardini, Cacciapuoti, Ferretti and Wick, who found a value of L much larger than that determined by other authors for nearly the same value of p_{min}. The reason for the disagreement is not completely clear. It may be noted that Ageno and collaborators used a lead absorber placed *between* the counters to compensate for the difference in atmospheric depth between the higher and the lower station. With this arrangement an appreciable number of mesotrons may have been removed from the beam by scattering, and this may have reduced the magnitude of the absorption anomaly due to decay. We may add that the results recently obtained by Neher and Stever[4] with an ionization chamber, concerning mesotrons coming in all directions, agree better with our present results and with those of Rossi, Hilberry and Hoag and of Nielsen, Ryerson, Nordheim and Morgan than with the results of Ageno, Bernardini, Cacciapuoti, Ferretti and Wick.

CONCLUSION

The experiments described have shown, in agreement with previous results, that the number of cosmic-ray mesotrons is more strongly reduced by a layer of air than by a dense absorber equivalent to the air layer with regard to ionization losses. The indication from earlier experiments[13] that the difference in stopping power between air and condensed materials increases when the mesotron momentum is decreased has been definitely established. This result verifies a theoretical prediction based upon the disintegration hypothesis, thus confirming the view that the absorption anomaly is caused by spontaneous decay of mesotrons in the atmosphere. A value of the proper lifetime $\tau_0 = 2.4 \times 10^{-6}$ sec. is deduced from measurements on a fairly monokinetic group of mesotrons.

We are greatly indebted to Professor A. H. Compton for discussions of the problem and for the encouragement given to this work. We also express our appreciation to Professor J. C. Stearns of Denver University for his friendly collaboration, and to Professor N. Hilberry and Mrs. Jane E. Hamilton for their valuable assistance throughout the experiments. To the Willet Company for making available for two months a suitable truck, to the Denver City Parks for their helpful cooperation and to the Carnegie Institution of Washington for financial support, we wish to express our sincere gratitude. One of us (B. R.) acknowledges with thanks the financial support granted to him by the Committee in aid of Displaced Foreign Scholars.

[13] See reference 2. Also M. A. Pomerantz, Phys. Rev. 57, 3 (1940).

THE SPACE TRAVELLER'S YOUTH

H. BONDI

Professor of Applied Mathematics, King's College, London

In recent times there has been a good deal of discussion of what people call the clock paradox of relativity. It may briefly be stated in the following form: Two people are together on the Earth; one stays on the Earth, the other embarks upon high-speed space travels, then returns to the Earth and meets the first man. Has the traveller aged just as much as the stay-at-home, or has he aged by a different amount? The problem is essentially one of the comparison of different motions, and the part of physics that is principally concerned with this kind of question is the theory of relativity. This celebrated theory falls into two quite distinct parts, the special theory and the general theory. Brief reference will be made to the character of each part, and the extent to which we have to make use of it.

COMMON SENSE IN PHYSICS

The result (which is that the space traveller has aged less than the stay-at-home) is sometimes called unacceptable, and this requires a little discussion. The result of a theory, when applied to a particular problem, may be logically absurd and may contain internal contradictions. This would indicate that the theory has been applied to the sort of problem to which it should not and cannot be applied. In fact, no such trouble arises here, but many people dislike and disbelieve the result the theory produces because, as they say, it contradicts common sense. This is an argument so frequently brought against the theory of relativity that it deserves a little investigation.

What is common sense, and what is its use in physics? Common sense is simply the accumulated experience we have of dealing with objects and people we have come in contact with during our life. As these encounters are very frequent we have accumulated a large amount of such experience, and common sense is a splendid guide for experiences of this kind. One of the most important results of this experience is that over a wide range of objects mere scale does not matter; for example, a rolling pea will, but for the imperfections of its shape, move just like a rolling billiards ball. To some extent the motions of the molecules of a gas may be pictured by a dust storm, although the size of the particles is very different. Similarly, sitting in an aircraft going at three hundred miles an hour is not essentially different from sitting in a car and going at thirty miles an hour, or from being carried along in a sedan chair at three miles an hour. It is this irrelevancy of scale that has made it so useful for us to construct models to visualise phenomena of inconvenient size. Thus we use a glass bead to study the refraction phenomena occurring in minute water droplets that produce rainbows, and scale models in general are widely used in engineering.

In the 19th century it was thought that the main purpose of physics was to construct models of convenient size and speed and duration to account for all the phenomena in the universe. In this century, we have learned better. We have learned that, while scale and speed and duration do not matter over a wide range, they are not irrelevant over extremely large ranges. Thus, while three hundred miles an hour is not very different from three miles an hour, and the same holds for three thousand or even three million miles an hour, the same is not true when we get up to five hundred million miles an hour, that is, when we get close to the velocity of light. Then speed does matter. Similarly, atomic theory has taught us that electrons and the nuclei of atoms do not behave like very small billiards balls, but behave quite differently. There is a limit to modelling. We cannot construct working models of convenient size and speed for all phenomena. If the speeds are very high or the sizes very small, then the familiar world of the objects around us is no guide. In these fields, then, common sense is no use at all; it is an intruder that merely befogs the issue and misleads us by applying analogies with experiences that are not analogous. We should, therefore, not be surprised if, when speeds well out of our daily experience occur in problems, the answer should at first sight appear odd and disagree with our intuitive ideas of what to expect.

A LONG WAY ROUND

In the whole of physics one meets the distinction between two quite different classes of quantities. These may be called route-dependent and route-independent quantities. The distinction is familiar from ordinary life although it is not always appreciated. Suppose that two people set out to walk by different routes from one place over the hills to another place where they meet again. It is then clear that the mileages they have walked may well be different, but that the net gain in height (difference between height climbed and descended) must be the same for both—in fact it must be the difference in the heights of the starting and finishing places of their walks. This demonstrates the distinction between the two classes of quantities. The net gain in height in going from one place to another is independent of the route taken; it depends only on the heights of the starting and finishing points, being in fact their difference. The distance walked is a quantity of quite different nature; it depends on the route taken.

These two classes of quantities occur in the whole of physics. To give one or two examples: the change in potential energy on moving a mass (or charge) from one point to another in a gravitational or electrostatic field is independent of the route taken. The heat produced by friction in taking a spoon in a pot of honey from one place to another depends on the route taken and the speed with which the spoon has been taken over the route. And so it goes on. There are more highbrow

Reprinted from Discovery
(Dec. 1957 - Pages 505-510)
By Permission

expressions for this division into classes, among them conservative forces and non-conservative forces, irrotational flow and rotational flow, potential fields and non-potential fields and so on, but the split occurs throughout the whole body of physics.

We can appreciate whether a quantity is route-dependent or route-independent only if we can vary the route or, to put it more strongly, only if we can vary the route appreciably. If two towns are connected by a narrow causeway which is the only way of getting from one to the other, then the inhabitants of these towns who never go elsewhere will have little conception of the route-dependence of mileage. The question of the space traveller's age is essentially the question of the route-dependence of the lapse of time. It is a question whether the time lapse between two occasions will be found to be the same by two people following different routes from one to the other or not. If, as it will turn out, the route-dependence is essentially a speed-dependence, then people who always move slowly will not be able to appreciate the route-dependence of time, and this in fact is the situation. Because our daily experience is with velocities small compared with the speed of light, we tend to think of time as a route-independent quantity. It is only when we consider theories (and the evidence on which they are based) concerned with high speeds that the route-dependence of time will become clear, although it can never be intuitively obvious to us, being outside the range of daily experience.

It will be instructive to consider the route-dependence of mileage in more detail. Suppose two motorists drive from one place to another. One follows the straight route, the other a curved one. The straight route will be shorter, as is obvious, but the connexion between the extra length of the curved route and the corners is not so clear. Suppose our second motorist follows a route consisting of a number of straight segments joined by short, sharp corners (Fig. 1). Of course his route will be longer than the route of the motorist driving along the straight road. His route will be longer *because* it curves, but the actual length of the curves themselves is quite negligible compared with the extra length of his drive. In other words, and this is the essential point of the problem, although the extra length is *due* to the curves it does not lie *in* the curves. One other point may be mentioned here. If one is a passenger in either of the two cars one will notice without looking outside whether the car is turning corners or not. One will notice the acceleration when turning the corner. Accordingly, if the two passengers compare experiences afterwards, then, without referring to the view, the passenger in the second car would say: "I knew my driver took a long and circuitous route, the corners were just terrible."

RELATIVITY THEORY

One of the most fundamental and obvious concepts of physics is acceleration, that is, change of velocity. We know that we can feel perfectly happy and stand and balance in a smoothly running train just as though we were on firm ground. This state of affairs persists as long as the velocity of the train is constant in magnitude

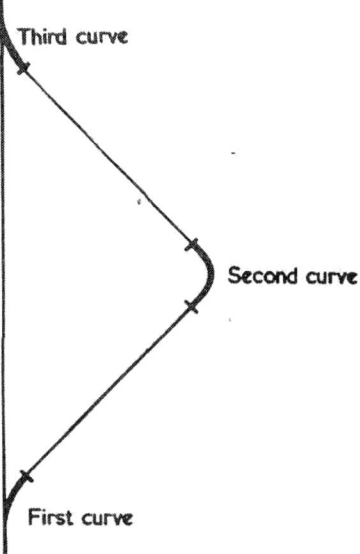

FIG. 1. Straight and circuitous travel by car.

and direction. But as soon as the train brakes sharply (or goes round a sharp bend) we have difficulty in balancing and are thrown around. Inanimate objects, like plates and glasses in the dining-car, are equally affected. As soon as there is a change of velocity—that is, an acceleration—its effects are plainly there, it can be felt. Thus there is a clear distinction between two modes of motion, accelerated and unaccelerated. Any person moving without acceleration is called an inertial observer. If an observer is inertial (unaccelerated) and a second observer is moving relative to him with uniform velocity in a constant direction, then evidently he too is unaccelerated and hence inertial. There is hence a whole family of inertial observers, moving relative to each other with uniform velocity in a constant direction, and this family is set apart from all other observers who experience accelerations and so are non-inertial. The principle of special relativity* now states that there are no distinctions within this family of unaccelerated observers. If any one of them carries out an experiment and gets a certain result, then any other one of them could carry out the same experiment and would get the same result. A non-inertial observer might, however, fare quite

* The fact that one speaks of the *theory* of relativity should not make the reader think that anything speculative is involved. The theory is so well confirmed by countless experiments and observations that it is one of the best established parts of physics. As there is no evidence against it at all, it would be unprofitable to doubt its predictions here.

differently. If an inertial observer releases a particle (not acted upon by forces) initially at rest relative to him, it will stay at rest relative to him. If a non-inertial observer does the same, then, as soon as he is accelerated, he and the particle will part company.

If one inertial observer makes himself a clock, driven by a clockspring, that ticks twice a second, then any other observer copying exactly this design of clock will have a device for measuring his own time in the same way, with two ticks a second. This is not a theorem but a definition, the definition of seconds for any observer. Instead of a spring-driven clock, any other device registering time could have been used, for example the spectral line of an atom, the time a specified acid takes to eat through a specified thickness of steel, the length of a generation of rabbits on a rabbit farm, the decay of a radioactive element, the ageing of human beings, and so forth. (The problems of biological time-keeping are discussed on pp. 519-21 of this number.) Moreover, all these devices measure the *same* time when travelling with one and the same observer. If one inertial observer notes that on an average human beings travelling with *him* cease to grow 1000 million ticks of *his* spring-driven clock after they have been born, any other inertial observer would find, by virtue of the principle of relativity, that on an average human beings travelling with *him* would cease to grow 1000 million ticks of *his* spring-driven clock after they have been born.

We can now draw a diagram to represent our results (Fig. 2). Let an inertial observer, *A*, make a map of his experience by drawing time upwards, distance (in one direction only—we do not need the other dimensions of space for our discussion) horizontally. He will then represent himself by the line *AA* (being always at zero distance from himself). The marks on this line (his world line in the language of relativity) will then represent the ticks of his clock. Any other inertial observer (one moving with constant velocity relative to *A*) would then appear in *A*'s diagram in a straight line *BB* since he would cover equal distances in equal times. The steeper the line *BB*, the more slowly *B* moves relative to *A*, since the smaller the distances he covers in each period of time. Light travels very fast and so light rays are represented by rather flatter lines (drawn dashed in Fig. 2).

Another inertial observer, *C*, may be in a different position from *A*, but without relative velocity, and hence, keeping the same distance, he will be represented by a line *CC* parallel to *AA*. Suppose now that *A* sends out a flash of light each time his clock ticks (1, 2, 3 . . .) rather in the manner of a lighthouse. The flashes are received by *B* at 1', 2', 3' . . . respectively, and by *C* at 1", 2", 3". . . . . If *C* compares the intervals 1", 2", etc., with the ticks of his own clock (constructed to be identical with *A*'s clock) then it is fairly evident that these will agree, since light takes just as long to get from 1 to 1" as from 2 to 2", the distance *AC* being constant. On the other hand, the time taken to get from 2 to 2' will be greater than the time taken to get from 1 to 1' since the distance 2,2' exceeds the distance 1,1'. Accordingly, if *B* compares the interval 1',2' with the ticks of

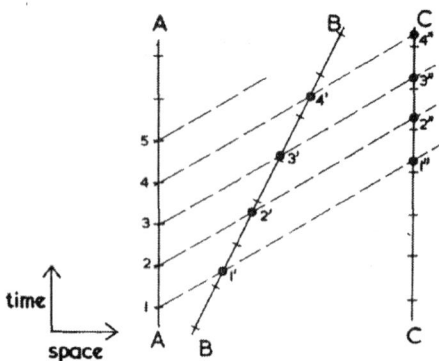

his clock (again identical with *A*'s clock), he will not get identity but will find the interval rather longer, say *k* times longer. How large *k* will be depends on his velocity of recession *v* relative to *A*, that is, it will be a function of *v*. Since *B* and *A* are identical observers* it follows that if *B* sent out flashes of light at each tick of his (*B*'s) clock, *A* would receive them at intervals equal in length to *k* times the interval between the ticks of *A*'s clock.

Moreover, it is clear that if *B* sends out flashes at 1', 2', . . . , *C* will receive them at 1", 2", etc. This means that if *B* sends out flashes at intervals equal to *k* times a clock tick, they will be received by *C* at clock tick intervals, that is, the interval is now multiplied by $1/k$. The velocity of *C* relative to *B* is evidently a velocity of approach *v*. Therefore, if a factor *k* corresponds to a velocity of recession *v*, the factor $1/k$ will correspond to the velocity of approach *v*.

The factor *k* is known from experiment as well as from theory and is greater than unity for a velocity of recession. It gives rise to the well-known Doppler shift of light which, in the case of recession, is a shift to the red, the period of the light wave being lengthened.

We are now in a position to analyse the clock paradox in the modified form due to Lord Halsbury (Fig. 3). An inertial observer *AA* is passed at *X* by an inertial observer *BB* travelling at velocity *v* relative to him and at a later instant *Z* by another inertial observer *CC* travelling relative to *A* with velocity *v* in the opposite direction to *B*. The two observers *B* and *C* pass each other at *Y*. They all carry identical clocks. At *X*, *A* and *B* both set their clocks to read zero hour. At *Y*, *C* sets his clock to read the same as *B*'s does at that instant. Does *C*'s clock show the same time at *Z* as *A*'s clock or not?

* This is the characteristically relativistic argument. Pre-relativistically the motion of the source and the observer relative to the "ether" would have to be known. The factor *k* would depend on whether the source was moving, with the observer at rest, or the source was at rest, with the observer moving.

To analyse this, suppose that A's clock shows $2t_0$ at Z and that B's clock shows t_1 at Y. Since B and C are symmetrically placed observers, the interval YZ will be the same on C's clock as XY is on B's clock, that is, it will be t_1, so that C will read $2t_1$ at Z. The question is then the relation between t_0 and t_1.

Let Q be the mid-point between X and Z which is of course on A's world line. Since the velocity of B relative to A is v, the distance of Y from A's world line (that is, from Q) in A's reckoning is vt_0, since the period XQ is t_0. Consider now a light ray emanating from Y. It will reach A at an instant Y' such that QY' equals the distance QY divided by the velocity of light c. Hence $QY'=vt_0/c$. Evidently Y' is the moment at which A sees Y happening.

Let h be the interval between the successive ticks of any of the identical clocks carried by A, B, C. The number of ticks of A's clock between X and Z is hence $2t_0/h$. The number of ticks of B's clock between X and Y is t_1/h. If at each tick B emits a flash of light, the interval between the reception of these flashes by A will be kh. The first of these flashes will be received at X, the last at Y'. Hence in the time XY' observer A receives t_1/h flashes at intervals of kh. Accordingly

$$t_0 + v\frac{t_0}{c} = t_0\left(1 + \frac{v}{c}\right) = kh \times \frac{t_1}{h} = kt_1. \quad . \quad . \quad (1)$$

Now consider C. His clock will tick t_1/h times between Y and Z. If at each tick a flash is emitted, the interval between the reception of successive ticks by A will be h/k, by what has been discussed with the aid of Fig 2. The first of these flashes will be received by A at Y', the last at Z. Hence in the time $Y'Z$ observer A receives t_1/h flashes at intervals h/k. Hence

$$t_0 - v\frac{t_0}{c} = t_0\left(1 - \frac{v}{c}\right) = \frac{h}{k} \times \frac{t_1}{h} = \frac{t_1}{k}. \quad . \quad . \quad (2)$$

Multiplying equations (1) and (2) we obtain

$$t_0^2\left(1 - \frac{v^2}{c^2}\right) = t_1^2, \quad . \quad . \quad . \quad (3)$$

whereas by dividing them we find

$$\frac{1 + \frac{v}{c}}{1 - \frac{v}{c}} = k^2. \quad . \quad . \quad . \quad (4)$$

Formula (4) is a well-known result for the Doppler-shift, amply confirmed by experiment, whereas (3) gives the answer to our problem, namely

$$2t_1 = 2t_0 \sqrt{1 - \frac{v^2}{c^2}}. \quad . \quad . \quad . \quad (5)$$

Therefore the clock readings at Z are not identical but the reading of C's clock is behind the reading of A's clock. The difference in the readings, $2t_0 - 2t_1$, depends very much on the velocity v. Table I gives some results for this difference if $2t_0$ is taken to be 20 years.[*]

TABLE I

v (miles/second)	v/c	Time difference	Farthest distance attained
1000	0·0054	2 hrs 32 min	20 light days
10000	0·054	10·6 days	0·54 light yr
93000	0·50	2 yrs 8 mths	5 light yrs (a little beyond the nearest fixed star)
149000	0·80	8 yrs	8 light yrs

ACCELERATION

In Lord Halsbury's formulation of the clock paradox *three unaccelerated* observers are used, whereas in the original formulation only two observers were used, at least one of them having to undergo accelerations. In order to consider the original problem we have hence to consider the effects of acceleration.

First, however, a common misunderstanding has to be removed. As has been said the *special* theory of relativity affirms the physical equivalence of all *inertial* observers. The *general* theory has occasionally been misunderstood to assert the equivalence of *all* observers, accelerated or not. This is, however, not so. The general theory only asserts

    (i) the *local* equivalence of the physical effects of acceleration and gravitation;

    (ii) the need for a mathematical formulation of the laws of nature identical in *form* (though of course not in content) for inertial and accelerated observers.

As for (i), the theory immediately supplies means of distinguishing between the effects of gravitation and acceleration when *extended* regions are considered, whereas (ii) is a purely mathematical demand of no direct physical significance.

In any case, it is obvious that no theory denying the observability of acceleration could survive a car ride on a bumpy road.

Secondly, we have to consider how accelerations will show up on our space-time diagrams. Since velocity appears as slope, acceleration—that is, change of velocity—will show up as change of slope and hence as curvature. Any curved part of a world line is an accelerated one, and the sharper the bend, that is, the higher the curvature, the greater the acceleration.

Thirdly, we have to think about the effect of acceleration on clocks. As long as the motion is unaccelerated the special theory of relativity tells us quite unambiguously that it does not matter what sort of a clock is

[*] Observer A should of course be unaccelerated whereas we have in mind using the Earth as A's station. The accelerations of the Earth are small, but would affect the results given by a few seconds. The effect is important for low-speed travel (30 miles per second or so) in the solar system and, together with other effects due to gravitation, requires general relativity for its treatment.

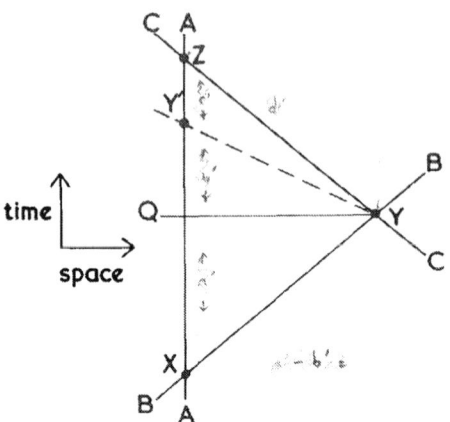

FIG. 3. The clock paradox in modified form.

used. It may be a wrist-watch or a vibrating atom, it may be the ageing of a human being or the decay of a radioactive nucleus. Every one of these, and all other methods of measuring time, will give the same result. But if there are accelerations what can we use then? This is an important question.

We know from daily experience that different clocks react differently to accelerations. A sensitive watch dropped on a concrete floor from the height of a few inches will stop working, whereas a shock-proof watch will go on working quite happily. A human being jumping up and down for a few feet of height will go on recording time by the ageing of his body cells and by the consciousness of his brain. A human being subjected to much larger accelerations will die and will cease to measure time. A vibrating atom can indicate quite large accelerations without being affected, but it is possible to shoot it to pieces in a big atom-smashing machine, and so there are limitations to its use too. Similarly, of

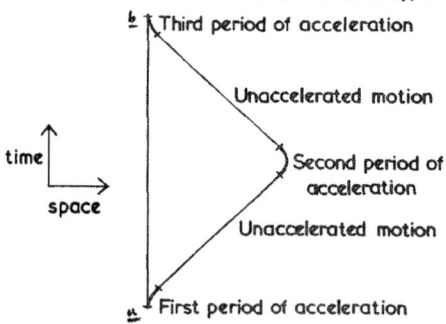

FIG. 4. Motion with short periods of high acceleration.

course, there were limitations in our earlier example to the sharpness of the corners a car could take. If they were too sharp it would overturn and that would be the end of that journey. So we must be a little careful about the use of watches and clocks in accelerated systems, but with proper limitation of their acceleration we can go on using them. We know indeed from our experience on the earth that jumping does not noticeably age us, whatever some people may say about the unhealthiness of exercise, nor does it stop our watches working or indeed falsify their readings. With a little care, therefore, this difficulty can be overcome.

Now, let us go back to the original question, and look at the space-time diagram of the earth-bound man and the space traveller (Fig. 3). The earth-bound man describes a straight path[*] (neglecting the Earth's rotation, etc.). The space traveller, since he follows a different route, must follow a curved path, that is, an accelerated path. If the space traveller follows a path with short periods of high acceleration, then his path in Fig. 4 will be very close to XYZ in Fig. 3, but we must be careful about the effect of the accelerations on his clocks and on his body. If he can stand high accelerations, then his path will be so close to that in Fig. 3 that we can use the results of our work (equation (5) and Table I). The space traveller will not have aged as much as the stay-at-home. Note that although the total time during which he was accelerated was short, yet as a consequence of these accelerations his whole lapse of time is affected, just as in the case of the driver's circuitous route in our analogy before.

With these considerations the discussion of our main problem has been completed. However, in view of the accelerations required, the applicability of our results to clocks of a delicate nature like human beings is still in doubt. To take an example, if accelerations of 20g were used (an acceleration in which every object presses on its support with 20 times its normal weight) for a period of just over 10 days, a speed equal to half the velocity of light would be attained. In the example of Table I (20 years' absence of the traveller) periods of acceleration of 10 days each are so short in comparison with the length of the journey that our work there can be used, and so we find a time difference of 2 years 8 months. A radioactive material would not be affected at all by such an acceleration, and its decay would show the effect clearly. A human being, on the other hand, would die. Can we devise a mode of acceleration that will be tolerable to humans? If we want to be careful about our space traveller's health, then we must not subject him to violent accelerations. He can reach high speeds by suffering accelerations of only very moderate magnitude for a long time. Technically, it is impossible to construct such a space-ship, but here we are not concerned with engineering possibilities but with biology. We can, then, imagine a space-ship whose acceleration is always thirty-two feet per second[2].[†] Then the space

[*] Note that there is only *one* straight path between two events. All other paths are necessarily curved.
[†] For short periods when near the Earth's surface somewhat higher, but wholly tolerable, accelerations would have to be endured.

this all uses light as a communicative medium
for continual comparison
what about restricting due relativity to a and b

TABLE II

A	B	C	D	E
36 days	nearly 36 days	5 mins 20 secs	Pluto (38·34 AU)	0·025
3 mths	nearly 3 mths	1½ hrs	1½ light days	0·064
1 yr	nearly 1 yr	2 days 16 hrs	23 light days	0·25
4 yrs	3 yrs 6 mths	6 mths	0·85 light yrs	0·72
40 yrs	11 yrs 9 mths	28 yrs 3 mths	18 light yrs	0·995
400 yrs	20 yrs 8 mths	379 yrs 4 mths	198 light yrs	0·99995
4000 yrs	29 yrs 7 mths	3970 yrs 5 mths	1998 light yrs	0·9999995
40000 yrs	38 yrs 7 mths	39961 yrs 5 mths	19998 light yrs (nucleus of our galaxy)	0·999999995

A: Time between departure and arrival of space traveller measured by a clock on the Earth.

B: Time between departure and arrival of space traveller measured by the space traveller's clock.

C: Difference A−B.

D: Farthest distance reached.

E: Ratio of highest speed attained to the velocity of light.

The equations from which the figures are derived are:

$$\frac{fs}{c} = \log_e\left[\frac{ft}{c} + \left(1 + \frac{f^2t^2}{c^2}\right)^{\frac{1}{2}}\right],$$

$$x = c\left[\left(t^2 + \frac{c^2}{f^2}\right)^{\frac{1}{2}} - \frac{c}{f}\right],$$

$$E = t\left(t^2 + \frac{c^2}{f^2}\right)^{-\frac{1}{2}},$$

where f = acceleration, c = velocity of light, $t = \frac{1}{2}A$, $s = \frac{1}{2}B$, $x = \frac{1}{2}C$.

traveller will feel very much at home inside the space-ship because conditions there will be identical with conditions on the Earth in every way, acceleration taking the place of gravity. Proceeding in this manner for a few years and then reversing the direction of accelera-tion and later on reversing it again, he can go very far indeed and come back, and yet suffer unhealthy conditions that might affect his rate of ageing (Fig. 5). It is true that he will feel a little giddy on the two occasions when the acceleration changes direction but one would not expect space travel to be the one form of travelling free of travel sickness. The theory of rela-tivity can be used to work out the time lapse recorded by the traveller and other details of his journey, as shown in Table II.

Whatever the mode of acceleration of the space traveller, whether short and violent or long and mild, he will know that he is being accelerated by the evidence of his senses. It will, therefore, not come as a surprise to him on his return to the Earth to find out that he has aged less than the people there, just as the traveller who took the curvy road cannot have been surprised that he covered a longer mileage than the traveller who followed the straight one. Hence there is no clock "paradox", since it is not paradoxical for two persons with different experiences to find that the consequences of their experiences differ. There is simply the result that high-speed travel makes the route dependence of time reckon-ing evident, whereas low-speed travel does not.

REFERENCES

There are several discussions of the "clock paradox" in

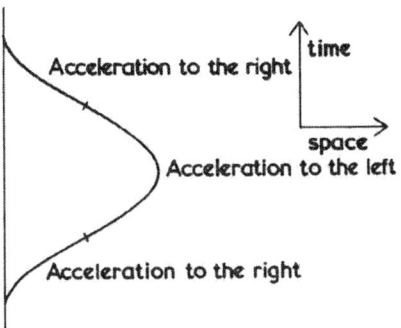

FIG. 5. Acceleration of constant magnitude with changes of direction. (A motion with constant acceleration is represented by part of a hyperbola.)

the literature, the most recent being an argument between H. Dingle and Prof. W. H. McCrea in *Nature*, vols. 177, 178, 179, in which Dingle takes an erroneous view based on the misinterpretation of general relativity referred to in this article. Lord Halsbury's formulation is given in one of Dingle's articles. Dr Crawford considers examples from physics in the same series, and acceleration is specifically considered by Sir Ronald Fisher and Prof. W. H. McCrea in DISCOVERY, vol. 18, p. 56 (1957).

THE RESOLUTION OF THE CLOCK PARADOX*

GEOFFREY BUILDER

University of Sydney, Sydney, Australia

Two ideal standard clocks, effectively isolated from interaction with other physical systems, and in a region of the universe free of gravitational fields, are assumed to move in any arbitrary manner so that they coincide on at least two occasions. In general, the reading of one of them will become retarded relative to the other in the interval between successive coincidences. This *relative retardation* is predicted by the restricted theory of relativity, taken together with the assumption that the 'rate' of a clock depends only on its velocity and not on its acceleration. The recognized procedure for calculating the retardation in terms of clock 'rates' is set out, and is illustrated by its application to a simple hypothetical experiment in which one clock remains at rest and the other travels away from and back to it with constant speed in a straight line.

There is nothing paradoxical in the predicted retardation as such. The so-called 'clock paradox' arises because an alternative, and apparently valid, calculation procedure, also based on clock 'rates', leads to a contradictory result. The term 'paradox' is something of a euphemism since the two predictions are contradictories.

It is shown that the paradox does not arise when direct use is made of the Lorentz transformation without introducing the additional, and non-essential, step in reasoning involved in utilizing the clock 'rates'. It is then shown that the paradox arises only through using these clock 'rates' without due regard to the exact significance of the quantities so described; once this is recognized the paradox is resolved completely within the framework of the restricted theory, which then provides a unique and unambiguous prediction of the relative retardation.

Once the paradox is thus resolved, the general theory of relativity can add nothing significant to the analysis. It is shown that the application of the principle of equivalence is essentially trivial; in effect, Einstein and Tolman evaded the real logical issue raised by the contradictory predictions by denying the applicability of the restricted theory and then utilizing, by means of the principle of equivalence, results obtained from it. This tortuous procedure succeeded in evading the paradox rather than in resolving it; it would obviously be quite invalid were the restricted theory indeed inapplicable to the problem.

1. **Introduction.** We consider two ideal standard clocks, effectively isolated from interaction with other physical systems or objects, in a region of the universe free of gravitational fields. The clocks may be supposed to move in any arbitrary manner whatsoever, subject to the condition that they coincide on at least two occasions. In general, the reading of one of the clocks will become retarded relative to the reading of the other in the interval between their successive coincidences. This effect may be referred to as the *relative retardation of clocks;* it is a prediction of the restricted theory of relativity taken together with the generally-accepted assumption that the "rate" of a moving clock at any instant depends only on the speed, and not on the acceleration, of the clock at that instant.

This assumption, and the corresponding assumption about the dimensions of bodies in non-uniform motion, cannot be justified by means of the restricted

* Received February, 1958.

Reprinted from Phil. Sci.
(April 1959 - Pages 135-144)
By Permission

theory. Their justification is essentially empirical; on the one hand, predictions based on them have never been found inconsistent with empirical data and, on the other hand, they are consistent with all our physical knowledge[1] about the properties and behaviour of bodies and clocks. These assumptions are generally regarded as being sufficiently well established to provide, when taken together with the restricted theory and the principle of equivalence, a satisfactory postulatory basis for the general theory of relativity.

The *recognized procedure* for calculating the relative retardation may be set out in the following steps:

(a) The motions of the clocks may be ascertained, or specified, in any one arbitrarily selected inertial reference system S.

(b) The Lorentz transformation can be used to predict that an ideal standard clock in motion with instantaneous speed v relative to the system S will, according to the measures of that system, run at a "rate" less than that of identical clocks at rest in that system by a factor $\sqrt{1 - v^2/c^2}$, where c is the velocity of light. [See equation (1) of section 3 below].

(c) Since the two clocks coincide on at least two occasions, their "rates" at each instant permit calculation of the time elapsed on each clock during the interval between the two events which are their successive coincidences.

(d) The elapsed time read on a clock between two events coincident with it is referred to as its *proper time* between these two events. It can be shown that the calculated value of this proper time is an invariant which is independent of the choice of the inertial reference system to which calculations and measurements are referred. That this must be so is obvious on general physical grounds, for *all* observers, irrespective of their state of motion, must agree as to the reading on a clock of the time of occurrence of an event coincident with the clock.

(e) In general, the *proper times* of the two clocks, between the two events which are their mutual coincidences, will be different since, in general, their motions relative to the arbitrarily selected inertial reference system S will be different. Thus one of the clocks must, in general, become retarded relative to the other in the interval between their successive coincidences, and the calculated relative retardation must itself be an invariant.

This method of calculation is valid whatever the motions of the clocks, so long as the "rate" $\sqrt{1 - v^2/c^2}$ of each clock is taken to correspond at each instant to the value v of its speed at that instant. The method can however be illustrated, without any significant loss of generality, by considering a simple hypothetical experiment.

----

[1] An investigation, to be published elsewhere, has shown that acceleration would have some effect even on ideal "relativistically-rigid" clocks but that the effect is not important; for real clocks the effect would be very much greater but could be taken into account in an actual experiment.

It is supposed that two ideal standard clocks R and M are originally at rest together at the origin of coordinates O of an inertial reference system S and that both are synchronized with the ideal standard clocks of this system. At the instant $t = 0$ read on all these clocks, the clock M suddenly moves away from R along the x-axis of S with the constant speed v. Its motion continues for a time T, as read on the clocks of system S, and is then suddenly reversed so that it travels back towards R again with the same speed v and reaches R at the S-time 2T. Since R remains one of the S-system clocks it also will read 2T when the clock M returns to it.

It will be supposed that the times required to accelerate M at the beginning of its outward and return journeys, and the times required to decelerate M at the end of each of these journeys, can be neglected without appreciable error. This can always be realized, for moderate accelerations, by supposing the time T to be very great (1), and it may also be justified for rapid accelerations (2).

During the whole of its outward journey, and during the whole of its return journey, M's speed measured in the system S is v and its 'rate' has the value $\sqrt{1 - v^2/c^2}$. Hence, neglecting the times required for acceleration and deceleration, the time read on it when it again coincides with R will be $2T' = 2T\sqrt{1 - v^2/c^2}$, so that it will then be retarded relative to R by the amount $2T - 2T\sqrt{1 - v^2/c^2}$.

There is nothing paradoxical in this predicted retardation as such. As Lovejoy (3) pointed out, there is no commonsense reason for doubting that the high speeds contemplated might affect the running of a clock or the senescence of an individual.

The paradox arises in the following way. During its outward journey from R, M is in uniform motion with speed v along the x-axis of system S, and is therefore at rest in a hypothetical inertial reference system S' moving with the speed v along the x-axis of S. According to the restricted theory, measurements made in the system S' would show that the clock R, which is in motion with speed v relative to S', has a "rate" of $\sqrt{1 - v^2/c^2}$ compared with clocks, such as M, at rest in S'. [See equation (2) of section (3) below.] Similarly, during its return journey M will be at rest in an inertial reference system S'' relative to which the clock R also moves with speed v, so that measurements made in S'' must also show that the "rate" of R is $\sqrt{1 - v^2/c^2}$ compared with clocks, such as M, at rest in S''. Thus these predicted measurements in S' and S'' appear to lead inevitably to the conclusion that M must become advanced relative to R during their separation, in direct contradiction to the previous prediction. This is referred to as the *clock paradox* of the restricted theory.

The term "paradox" is something of a euphemism since the two predictions are if fact contradictories.[2] As such they have been a powerful weapon for

---

[2] It has sometimes been inferred that both predictions must be wrong and that therefore the relative retardation must be zero; but such a conclusion is quite incompatible with the restricted

opponents of the restricted theory and a serious embarassment to its adherents, Nevertheless, the theory survived simply because of its spectacular success in all other respects.

The real logical issue was finally evaded by Einstein(4), Tolman(5) and others by denying the applicability of the restricted theory to the problem. No rigorous justification for this denial was ever given; but the evasion was given a degree of plausibility by the claim that the general theory of relativity resolved the paradox and confirmed the validity of the steps (a) to (e) of the recognized procedure of·calculation given above.

The current attitude which accepts this point of view has been stated by Professor Grünbaum in a recent paper in this journal (6), in which he has also set out a succinct account of the history of the paradox and has given references to a number of the more important papers and discussions relating to it. Yet it seems clear that, in spite of his stated acceptance of the current attitude, Professor Grünbaum is still troubled by a justified feeling of dissatisfaction and has consequently, like Professor McCrea (1), endeavoured to show that there really is no paradox even in the context of the restricted theory. But he did not succeed in doing more than demonstrating the invariance of the calculated retardation referred to in steps (d) and (e) of the recognized procedure outlined above.

In contradiction to the currently accepted view of the problem, I have recently shown (7) that the paradox can be resolved completely within the context of the restricted theory. The seeming paradox can, and does, arise only through using the "rate" of a moving clock without due regard to the precise definition of the quantity so described. Properly applied, the restricted theory gives a unique and unambiguous value for the relative retardation of the two clocks. The general theory of relativity can then contribute nothing further of physical significance to the analysis of the problem. In particular, any use of the principle of equivalence is trivial except in cases, lying outside the scope of the stated problem, in which a real gravitational field is present and must be taken into account as, for example, in the case discussed by Mikhail (8).

2. The Lorentz Transformations. As stated above, the paradox arises from a calculation based on the "rate" of the clock R as measured in the inertial reference systems S' and S''. This "rate" had been derived by means of the restricted theory i.e., by using the Lorentz transformations.

Such calculations constitute an indirect application of the Lorentz transformation to the problem. In spite of its apparent convenience, this indirect method is by no means necessary and it has the obvious disadvantage of introducing an additional and non-essential step in reasoning. That this additional step could be a possible source of fallacy is suggested by the following considerations.

theory. The only legitimate conclusion is either that the restricted theory is invalid or, as demonstrated here, that one of the predictions is fallacious.

According to the restricted theory, the Lorentz transformations relate the place and time (spatial and temporal coordinates) of the occurrence *of an event* in any one inertial reference system S to the place and time of occurrence *of the same event* in any second inertial reference system S'. Thus, in the context of the restricted theory, the transformations are directly applicable only to specified identifiable events. Yet in the calculation leading to the clock paradox, outlined above, there was no reference to any such events; the calculations were based solely on a statement of the behaviour of the clocks R and M in terms of the "rates" of these clocks while they were in uniform motion relative to one or other of the inertial reference systems considered.

The question therefore arises as to how the identifiable events, that would have been necessary to the direct application of the Lorentz transformation to the problem, have been eliminated from the discussion. It must also be asked whether this elimination has resulted in a loss of essential data and thereby permitted the paradox to arise.

It is therefore necessary to ascertain what conclusions are reached when the Lorentz transformations are applied directly to the problem. To do this, it is necessary to specify the identifiable events in the hypothetical experiment being considered. There are three such events:—

E_1 The departure of M from R.
E_{23} The reversal of M's motion as the end of its journey outward from R.
E_4 The arrival of M again at R.

All these events coincide with M. Thus the events E_1, E_{23} occur at the same place in the inertial reference system S' in which M is at rest during its outward journey, and the events E_{23} and E_4 occur at the same place in the system S'' in which M is at rest during its return journey. On the other hand, only the events E_1 and E_4 coincide with R, and only these occur at the same place in the system S.

I have shown elsewhere (7) that by utilizing the Lorentz transformations directly to determine the time intervals between these events, as measured in the systems S, S' and S'', and hence by the clocks R and M, one obtains the unique and unambiguous result that the clock M must become retarded relative to the clock R by the amount $2T - 2T\sqrt{1 - v^2/c^2}$ just as calculated by the recognized procedure given in section (1) above.

It is not possible, in this direct application of the transformations, to get any different result by any variation or twisting of the details of the calculations. For example, precisely the same result is obtained if one starts by specifying the duration of M's journey as being 2T', as read on M, and then calculates from this interval, read on the clock R, between the beginning and end of M's journey.

Thus it is clear that the paradox can, and does, only arise through the introduction of the additional step in calculation, in which the "rate" of a moving clock is utilized, and through the loss of some essential data in this additional step.

The "Rate" of a Moving Clock. The term *rate* is one which has long been in everyday use in reference to the speed at which a process takes place, as compared with some implied or stated standard. In particular, in non-relativistic physics, one is used to speaking of the rate of a clock, either in comparison with another clock or, more frequently, in comparison with an ideal clock which keeps a standard terrestrial time.

In this usage, the *rate* of a clock is essentially a measure of its departure from an ideal.

However, if the clock is a good one, it is commonly true that, for practical purposes, the *rate* can be taken to be constant over reasonably long periods and can be taken to be independent of the range of conditions to which such a clock is normally subjected, such as motion in a ship or aeroplane. Moreover, we are used to the practice of comparing such a clock, whether it is in motion or not, with our standard of time by means of radio time-signals and, if necessary, making a correction for the time of transmission of these signals from their source. Thus we are used to the idea that if we know the *rate* of such a clock we can calculate immediately how much its readings will diverge from the standard after any stated interval of time, without considering the motion of the clock or its position on the earth.

This term (or the concept to which it corresponds) has been taken over into the context of the restricted theory of relativity but with a significant change of connotation which we here indicate by the use of quotation marks.

This "rate" can no longer be a measure of the divergence of a clock from an ideal, for in this context we deal only with ideal standard clocks all of which are, by definition, identical and all of which behave in exactly the same way as each other when subject to the same conditions. They are ideal in the sense that they are supposed not to be affected by temperature and other physical conditions not specifically taken into account in the particular context.

The "rate" is used solely to refer to the effect, as measured in an inertial reference system, of the motion of such an ideal standard clock relative to that system. The clocks used for measurements in such a system are themselves identical ideal standard clocks at rest at various points in the system. Now if the observers in such a system S set out to determine the "rate" of a clock M moving relative to their system, any method they can devise turns out to be equivalent to one basic procedure; they can, in effect, only compare the readings of this clock M with the readings of two spatially separated synchronized clocks C_1 and C_2 at rest in their own system.

Suppose that M is moving with speed v along the x-axis of S and that its readings are t' and $t'+dt'$ at the instants it coincides with two clocks C_1 and C_2 placed at points spatially separated along the x-axis of S. Let the readings of C_1 and C_2 be t and $t+dt$ respectively at the instants M coincides with them. Then the "rate" of the moving clock, compared with the clocks at rest in S, is given by the ratio dt'/dt.

To predict the result of this measurement, we consider the inertial reference system S' in which the clock M is, for the time being, at rest at some point x' on the x'-axis. The Lorentz transformations can then be applied to the two identifiable events which are the coincidences of the clock M with the clocks C_1 and C_2. We thus obtain (7) for the "rate" of the clock M:—

$$dt'/dt = \sqrt{1 - v^2/c^2} \qquad (x' \text{ constant}) \qquad (1)$$

where the restrictive condition x' *constant* specifies that the given value for dt'/dt is applicable only to two events at the same place, e.g., coincident with M, in S'.

Thus equation (1) *states that the clock M records an interval dt', between two events coincident with it, which is shorter, by a factor* $\sqrt{1 - v^2/c^2}$, *than the interval dt, between the same events, as recorded on the spatially-separated clocks in the system S.*

This is the precise meaning of the "rate" dt'/dt given by equation (1). It is therefore the precise meaning of the "rate" of the moving clock M as measured in the system S.

Thus this "rate" of the clock M can be used only in calculations relating to the interval between events coincident with the clock M, i.e. in calculations of the *proper time* of clock M between events coincident with it.

Similarly, if we suppose there to be observers in the inertial reference system S' in which M is, for the time being, at rest, and if these observers measure the "rate" of the clock R, which is at rest in S and moving relative to S' with speed v, we can predict that they will find

$$dt/dt' = \sqrt{1 - v^2/c^2} \qquad (x \text{ constant}) \qquad (2)$$

i.e. *that the clock R records an interval dt, between two events coincident with it, which is shorter, by the factor* $\sqrt{1 - v^2/c^2}$, *than the interval dt' between the same events as recorded on spatially separated clocks in the system S'.* This therefore is the precise meaning of the "rate" of the clock R as measured in S'.

Thus the precise meaning of the "rate" of a moving clock as measured in an inertial reference system involves an essential reference to identifiable events. Such "rates" can therefore, like the Lorentz transformations, only be used in calculations dealing with specified identifiable events. Once this is recognized, calculations utilizing these "rates" can be expected to give results identical with those obtained by direct application of the Lorentz transformation and therefore free of ambiguity and paradox.

4. The Resolution of the Paradox. The events E_1 and E_{23}, which mark the beginning and end of M's outward journey, occur at the same place in the system S' and both coincide with M. Equation (1) therefore gives correctly the relation between the interval between these events as read on M, i.e., M's proper time, and the interval between these same events as read on the clocks of system S and therefore on the clock R.

Similarly events E_{23} and E_4, which mark the beginning and end of M's return journey, occur at the same place in the system S″ and both coincide with M. Equation (1) therefore also gives correctly the relation between M's proper time and the interval between these events as read on the clocks of system S, and therefore on R. The retardation of M relative to R can therefore be calculated correctly using the recognized procedure set out in section (1), in which equation (1) is used.

On the other hand, since the events E_1 and E_{23} do not occur at the same place in system S, the "rate" of R's clock given by equation (2) cannot be applied to calculations of the interval between these events. This also applies to the interval between E_{23} and E_4 since these events also occur at different places in S.

Thus the calculation of the relative retardation of the clocks, using the "rate" of equation (2) and giving rise to the paradox, is fallacious and must therefore be rejected completely.

The paradox is thus resolved and is shown to have arisen out of lack of precise definition, and consequent misapplication, of the "rate" of a moving clock.

5. The Asymmetry of the Retardation. The paradox has sometimes been stated as a contradiction between the asymmetry of the predicted retardation and the symmetry between the "rate" of the clock R measured in S′ and the "rate" of the clock M measured in S. This contradiction ceases to have any meaning once it is recognized that the "rate" of clock R measured in S′ has no relevance whatsoever to the experiment considered[3].

The asymmetry of the predicted retardation is obviously consistent with the dynamical asymmetry involved in the experiment itself, which required that M should be subjected to accelerations while R remained at rest in an inertial reference system. This dynamical asymmetry is displayed in the calculations as an asymmetry of the relation of the identifiable events to the two clocks, i.e. all these events coincide with M but only the first and the last with R; it is therefore displayed as an asymmetry in the relation of the identifiable events to the reference systems S and S′ and to the systems S and S″.

We may therefore claim fairly, if somewhat picturesquely, that the restricted

[3] This "rate" of R, as measured in S′, can however be utilized if *all* the calculations and measurements are referred to the system S′. In this case the corresponding "rate" of M's clock at each instant, as measured in S′, must also be used. When this is done the calculated value of the relative retardation is, of course, identical with that previously found since the procedure complies with the statement of the recognized procedure of calculation set out in section (1), and its result illustrates the invariance of the calculated retardation. That the result obtained by calculations referred to the system S is identical with that obtained by calculations referred to the system S′ was demonstrated by Professor Grünbaum (6).

theory is not deceived by the purely kinematical symmetry of the motions of M and R relative to one another. In effect, it takes into account also the motion of each, relative to its own inertial systems of reference, as specified by the identifiable events.

6. **The General Theory of Relativity.** The general theory can add nothing of physical significance to the foregoing analysis. Once the paradox has been resolved within the context of the restricted theory, it is pointless to adduce the general theory at all. In particular, it can be shown that the application of the principle of equivalence of the general theory, as in the treatment given by Einstein and Tolman, is essentially trivial. The reasons for this can be shortly set out as follows.

The principle of equivalence states that the description of events in terms of the coordinates of an accelerated reference system is indistinguishable from the description of identical events in terms of the coordinates of a reference system at rest in an equivalent gravitational field. The principle thus permits the course of events in a gravitational field to be predicted by calculating, *by means of the restricted theory*, the course of events as described in terms of the coordinates of the equivalent accelerated reference system.

The converse process of calculating the course of events in terms of the coordinates of an accelerated reference system (such as a system in which M remains at rest throughout the whole of its journey to and from R) from the course of events in an equivalent gravitational field *must necessarily be trivial* since it would, in principle, involve first calculating the latter from the former.

In effect, Einstein (4) and Tolman (5) resolved the paradox by denying the applicability of the restricted theory, and then using instead conclusions that had been derived from the restricted theory by means of the principle of equivalence. This tortuous procedure succeeded in evading the paradox rather than resolving it. It need scarcely be pointed out that the procedure would be quite invalid if the restricted theory itself were indeed not properly applicable to the problem.

The use of the general theory was however successful in giving an unambiguous and unique value for the amount of the relative retardation. The reason for this success is now not difficult to see. The steps, starting from the Lorentz transformations, to accelerated reference systems and thence, by the principle of equivalence, to reference systems at rest in equivalent gravitational fields and thence, again by the principle of equivalence, back to the particular accelerated reference system in which M is at rest throughout his journey, were all made without any misapplication of the clock "rates" given by equations (1) and (2). The fallacious application of equation (2) was thus eliminated, so that the treatment was rigorous throughout even though it was extraordinarily cumbersome and even though it becomes essentially trivial once the paradox has been resolved within the context of the restricted theory itself.

REFERENCES

1. McCrea, W. H.: *Nature, 167,* 680 (1951).
2. Moller, C.: "The Theory of Relativity", Oxford: Clarendon Press, 1952.
3. Lovejoy, A. O.: *Philosophical Review, 40,* 48 (1931).
4. Einstein, A.: *Naturwissenschaften, 6,* 697 (1918).
5. Tolman, R. C.: "Relativity Thermodynamics and Cosmology", Oxford: Clarendon Press, 1934.
6. Grünbaum, A.: *Philosophy of Science, 21,* 249 (1954).
7. Builder, G.: *Aust. Journ. Phys., 10,* 246 (1957).
8. Mikhail, F. I.: *Proc. Camb. Phil. Soc., 48,* 608 (1952).

LETTERS TO THE EDITORS

The Editors do not hold themselves responsible for opinions expressed by their correspondents. No notice is taken of anonymous communications.

Experimental Verification of the 'Clock-Paradox' of Relativity

EINSTEIN's theory of relativity predicts that if one of twin brothers leaves home in a space ship and spends a considerable amount of time travelling at high velocity with respect to his home inertial frame, then, when the traveller returns home, he will find himself physiologically younger than his stay-at-home brother, and his pocket watch will have performed correspondingly fewer revolutions.

In order to guarantee the above prediction, three assumptions are sufficient : (1) The time dilation of special relativity holds for uniform motion. (2) The acceleration of an ideal clock relative to an inertial system has no influence on the rate of the clock, and the increase in the proper time of the clock at any time is the same as that of the standard clocks in the system in which the clock is momentarily at rest[1]. (3) The traveller and his pocket watch are good approximations to an ideal clock. (The accelerations must not kill the traveller or break his watch.)

The description of the trip from the point of view of the traveller's (non-inertial) rest frame is complicated by the accelerations. This complication has led historically to the name 'clock-paradox'. The 'paradox' can be 'resolved' by the use of general relativity[2][3]. (However, even to describe the trip from the traveller's accelerating frame, special relativity plus assumptions 2 and 3 are sufficient (H. Stapp, private communication).)

An experimental test would be highly desirable, because (*a*) there is not universal agreement that relativity actually does predict the asymmetrical ageing[4], and (*b*) assumption 2 does not follow from special relativity[1]. Verification of the 'clock-paradox result' would therefore require that assumption 2 be included in any correct generalization of special relativity.

In order to look for an asymmetrical ageing effect, it is not necessary for the traveller to make the usual round trip. He can make a one-way trip, stop at his destination, and then compare his age with the stay-at-home twin by means of radio signals. This follows from the basic assumption of special relativity that a common time can be defined everywhere in a given inertial frame (the home frame) by use of local clocks that have been synchronized by means of light signals[5]. The traveller can compare his pocket watch with a *local* clock at his destination, after he is at rest in the home frame. As an additional test, he could then return home at a comparative snail's pace, and compare his pocket watch directly with that of the stay-at-home. He would find that the age difference acquired during the fast outward trip is maintained unchanged during the return trip, provided that the traveller's return velocity is small enough. This follows from the fact that

$$(c/v)\,\{(1 - v^2/c^2)^{-1/2} - 1\} = 1/2(v/c) + \dots$$

tends to zero as $v/c \to 0$.

Mesons are clearly a suitable substitute for brothers[6]. A simple accelerator experiment would

use a uniform, parallel beam of mono-energetic π^+ mesons impinging on two identical detectors located at different distances from the target producing the mesons. Variation of the meson flux with distance from the target would check assumption 1. To check assumption 2, the mesons would be decelerated to rest at each detector, and the number of decays at rest compared to the number of incident pions.

The corresponding one-way trip differs from that mentioned above, since it begins with both brothers at the same place and at relative rest in a moving inertial frame, and ends with them both at relative rest but separated in a new inertial frame (the laboratory). This trip has the virtue that it is completely symmetrical in the treatment of the two brothers, *except that one of them is decelerated while they are together, the other when they are apart.* An argument commonly used in trying to make the asymmetrical ageing plausible is that the asymmetrical behaviour of the accelerometers (or stomachs) of the brothers can distinguish the one who travelled from the one who stayed home. In our experiment that argument breaks down completely, although it is easily shown that assumptions 1, 2 and 3 still predict asymmetrical ageing. (It turns out that whichever brother accelerates into the other's rest frame *while the brothers are separated* will remain the younger (and is therefore by definition the 'traveller'), irrespective of which brother was originally accelerated, when they were together, to produce the relative motion.)

This proposed experiment has already been performed implicitly. The first quantitative check of assumption 1 is contained in the combined experiments of Rossi, Hilberry and Hoag[7], Rasetti[8], and Blackett[9]. Rossi *et al.* measured the decrease in μ-meson flux between mountain altitudes and sea-level, and by measuring and correcting for that part of the loss that was due to ionization-energy loss of the mesons in the air-path, they arrived at a mean life-time for decay in flight of about 30×10^{-6} sec. From the measurements by Blackett on the momentum distribution of μ-mesons at sea-level, Rossi *et al.* deduced a mean value for $(1 - \beta^2)^{-1/2}$ of about 15, from which they predicted a mean decay life-time of about 2×10^{-6} sec. for μ-mesons at rest. Rasetti later measured the mean life of μ-mesons at rest and obtained $(1 \cdot 5 \pm 0 \cdot 3) \times 10^{-6}$ sec. Thus assumption 1 has been verified.

In the Rossi experiment, the flux is measured at two places, *without appreciable deceleration* of the mesons. But the 'twin paradox' is not even qualitatively discernible in any experiment that does not involve relative accelerations, for then the brothers never separate, or else never return. In order to complete the experiment, it is necessary to decelerate the μ-mesons and observe their decay at rest—once at mountain altitudes and once at sea-level. Given assumption 1, the only way in which Nature can avoid asymmetrical ageing is by causing the sea-level mesons to undergo anomalously large ageing during their deceleration to rest, such that when at rest they have the same age as the mesons which were stopped at the mountain absorber and have been decaying at the normal rate. The excess decays during deceleration lead to correspondingly fewer decays at rest. This peculiar behaviour cannot be ruled out by special relativity alone, but *is* ruled out if we add assumption 2. It is apparent that the same asymmetrical ageing will also result from a weaker assump-

Reprinted from Nature
(Jan. 1957 - Pages 35-36)
By Permission

tion than No. 2, namely, that any anomalous ageing due to acceleration alone is independent of the history of the particle. To verify assumption 2, we must show that *no* anomalous ageing occurs during acceleration.

This last necessary experiment has been performed. Harold Ticho observed the decay of positive μ-mesons that had been decelerated to rest. The experiment was performed[10] at 11,500 ft. and at 600 ft. (Chicago) with the same apparatus (H. Ticho, private communication). Fast incident muons triggered the counter system ; delayed counts due to radioactive decay at rest were then observed. From the known momentum spectrum of the mesons at sea-level and at mountain altitude and the calculated geometrical efficiencies, Ticho could predict the number of fast muons that should stop in the absorber at either place and register their decays. If there were no asymmetrical ageing, Ticho would have observed at Chicago a rate anomalously reduced by a factor of about 40. Instead, Ticho observed roughly the expected number of decays at rest, both at low altitude and at high altitude. Thus, assumption 2 has been verified.

I conclude that the experiment that we have proposed for testing the clock-paradox result has already been carried out, through the combined experiments of Rossi, Hilberry and Hoag[7], Rasetti[8], Blackett[9], and Ticho (ref. 10 and private communication). Their results verify the asymmetrical ageing predicted by Einstein's theory of relativity.

I am grateful to Drs. Stelpen Gasiorowicz, Joseph Lepore, Maurice Neumann, Wolfgang K. H. Panofsky and Henry P. Stapp for illuminating discussions. This work was done under the auspices of the U.S. Atomic Energy Commission.

FRANK S. CRAWFORD, JUN.

Radiation Laboratory,
University of California,
Berkeley, California.
Oct. 24.

[1] Møller, C., "The Theory of Relativity", 49 (Oxf. Univ. Press, London, 1952).
[2] Møller, C., "The Theory of Relativity", 258 (Oxf. Univ. Press, London, 1952).
[3] Tolman, R. C., "Relativity, Thermodynamics, and Cosmology", 194 (Oxf. Univ. Press, London, 1934).
[4] Dingle, H., *Nature*, 177, 782 (1956).
[5] Møller, C., "The Theory of Relativity", 31ff. (Oxf. Univ. Press, London, 1952).
[6] The possibility of using accelerator-produced π-mesons to verify the 'clock-paradox prediction' was pointed out by E. Martinelli and W. K. H. Panofsky, *Phys. Rev.*, 77, 465 (1950).
[7] Rossi, Hilberry and Hoag, *Phys. Rev.*, 57, 461 (1940). I thank Prof Luis W. Alvarez for directing my attention to the cosmic-ray μ-meson experiments.
[8] Rasetti, F., *Phys. Rev.*, 60, 198, (1941).
[9] Blackett, P. M. S., *Proc. Roy. Soc. A*, 159, 1 (1937).
[10] Ticho, H., *Phys. Rev.*, 72, 255 (1947).

Reprinted from Nature
(June 1957 - Pages 1242-1243)
By Permission

The 'Clock Paradox' of Relativity

DR. FRANK S. CRAWFORD's further communication[1] is welcome as the first attempt to answer my arguments. Hitherto they have been ignored, and independent reasons, which I reject, have been adduced for the opposite conclusion. That leads nowhere.

Dr. Crawford rightly distinguishes the "qualitative" from the "quantitative" argument : these are based respectively on the 'postulate of relativity' and the 'postulate of constant light-velocity', which Einstein sharply distinguished from one another. The first states that Nature provides no phenomenon corresponding to absolute rest (later extended to cover motion in general) ; it is a generalization from experience and is non-mathematical. The second gives a definition for timing distant events ; it is entirely mathematical and is chosen to ensure that violations of the postulate of relativity shall not be expected. Being a definition, it is, of course, not susceptible of proof ; it is suggested by experience and is tenable so long as its implications conform to experience—in particular, the experiences leading to the postulate of relativity.

The argument against asymmetrical ageing rests wholly on the postulate of relativity, though I maintain that the postulate of constant light-velocity is compatible with it : the postulate of constant light-velocity is necessary to determine the age at which the twins will reunite. The argument is a single syllogism :

(1) According to the postulate of relativity, if two bodies (for example, two identical clocks) separate and reunite, there is no observable phenomenon that will show in an absolute sense that one rather than the other has moved.

(2) If on reunion one clock were retarded *by a quantity depending on their relative motion*, and the other not, that phenomenon would show that the first had moved and not the second.

(3) Hence, if the postulate of relativity is true, the clocks must be retarded equally or not at all : in either case, their readings will agree on reunion if they agreed at separation.

No flaw has been found in this. Dr. Crawford speaks of a retardation arising from "third body acceleration" with respect to the rest of the universe, "not the relative acceleration of the two twins". That is not "a quantity depending on their relative motion". But the only retardation in question in this discussion, namely, $2T\alpha \equiv 2T(1 - V^2/c^2)^{1/2}$, where $2T$ is the time of the round trip and V is the uniform relative velocity[2], *is* such a quantity. Hence his criticism is irrelevant.

I have said more than once[3] that I do not dispute a possible slight effect associated with third-body acceleration, but I cannot see how that, regardless of its magnitude, can turn $2T\alpha$ into a physical effect on a twin. This would be equivalent to the conversion of the apparent diminution of size which each of two observers sees in the other when they walk away from one another, into an actual shrinking of one of them if he alone does the walking. No one has yet explained this, and I would be grateful if Dr. Crawford would do so. If it does occur, however, then since we know, from Foucault's pendulum, for example, that the Earth has third-body acceleration, the Earth twin also should suffer the retardation and so be indistinguishable from his brother. Moreover, since Dr. Crawford holds[4] that it is only the acceleration *"while the brothers are separated"* (that is, that which occurs after a time T) that produces this effect (thus implying an unexplained distinction among third-body accelerations), how can the effect be proportional to $2T$? It would have to act retrospectively and make the ageing brother suddenly grow younger again on reversal. Is this credible ?

Now consider the quantitative argument. If my syllogism holds, then such arguments are not determinations of the actuality of the asymmetry but tests of the postulate of constant light-velocity : if the postulate of constant light-velocity requires asymmetry, it fails to conform to the postulate of relativity and must be rejected. (An *experimental* proof of asymmetry, on the other hand, would require that the postulate of relativity must be rejected.) I do not believe this. Space will not allow a direct answer to Dr. Crawford's criticism of my theorem which he quotes, which I do not accept ; but in view of its importance (I am surprised to find that this property of the Lorentz transformation is so little known) I will restate it and give another and more general proof to which the criticism cannot apply.

The theorem is this. Let B_1 and C_1 be two observers-cum-clocks moving with the same velocity V away from and towards, respectively, a distant object A_1 regarded as stationary. Then the events on A_1 which B_1 and C_1 regard as simultaneous with their meeting are respectively before and after, by equal amounts, the event on A_1 which an observer on A_1 regards as simultaneous with that meeting. To prove this, introduce as an intermediary an observer-clock D_1, stationary with respect to A_1 and coincident with B_1 and C_1 at their meeting. Let all three clocks, B_1, C_1, D_1, read 0 at that event, so that their co-ordinates are related by the Lorentz transformation. If x is the distance between A_1 and D_1 in their common rest frame, the event, E_0, on A_1 which is simultaneous in the D_1 co-ordinate system with the meeting of B_1, C_1, D_1 is $(x,0)$ in that system. Hence, by the Lorentz transformation, its time in the B_1 system is $Vx/c^2\alpha$, and in the C_1 system, $- Vx/c^2\alpha$. Now let A_1 be a clock identical in working with B_1 and C_1 but set at random, so that its reading at the event E_0 is T_0, which may be anything at all. Then clearly the readings of A_1 which B_1 and C_1, respectively, regard as simultaneous with their meeting are $T_0 - Vx/c^2\alpha$ and $T_0 + Vx/c^2\alpha$. A_1 will obviously regard the meeting of B_1 and C_1 as occurring at T_0, for that is his reading for the event which D_1 times at zero. This proves the proposition. It is easily verified that $Vx/c^2\alpha = \frac{1}{2}(x/c\beta - x\beta/c)$ of the former proof.

Dr. Crawford has agreed that my reasoning, apart from this theorem, is correct. I hope he will now be able to accept my result. But if not, how will he explain the asymmetry in this case, in which no acceleration of any kind occurs, when he has relied on third-body acceleration alone to produce it in the ordinary case ?

I must add a word on Einstein's "regrettable mistake". I have no doubt that Einstein himself realized this. In a later discussion of the clock paradox[5], he stated clearly that the problem cannot be considered in terms of the special theory since it involves accelerations. That alone would require the rejection of the earlier argument. But about the same time he wrote a popular book, translated into English as "The Theory of Relativity" (Methuen, 1920), in which all the reasoning of the 1905 paper (other than the technical electro-magnetic part) is retraced in order—with one omission, namely, this phenomenon of the circulating clock. This could not be because of its abstruseness—it is the simplest possible application of the theory—for even aberration and the Doppler effect are explained. I can think of no explanation except that, as he said elsewhere, he then considered the special theory inapplicable to such a problem.

HERBERT DINGLE

Purley, Surrey.

[1] *Nature*, **179**, 1071 (1957).
[2] See McCrea, *Nature*, **167**, 680 (1951) ; *Discovery*, **18**, 175 (April 1957).
[3] For example, *Nature*, **178**, 680 (1956) ; *Bull. Inst. Phys.*, **7**, 320 (1956).
[4] *Nature*, **179**, 35 (1957).
[5] *Naturwiss.*, **6**, 697 (1918).

LETTERS TO THE EDITORS

The Editors do not hold themselves responsible for opinions expressed by their correspondents. No notice is taken of anonymous communications.

The 'Clock Paradox' of Relativity

It would be inappropriate to give here an additional derivation of the asymmetrical relative ageing of twin brothers predicted by relativity theory for the familiar round trip. In his first paper on relativity, A. Einstein derives the well-known result[1]; he does not, as claimed by Prof. H. Dingle, make a "regrettable error"[2]. The recent derivation by W. H. McCrea[3] emphasizes that the Lorentz transformation of special relativity is sufficient to describe the round trip from the point of view of either twin, provided that one retains the usual central role assigned by Einstein to "a system of co-ordinates in which the equations of Newtonian mechanics hold good (that is, to the first approximation)"[4].

In view of the recent prolonged controversy[5], a more fruitful approach might be to find the specific source of the "regrettable error" on Prof. Dingle's part. I have done this, by examination of Prof. Dingle's recent Physical Society paper[6]. There (p. 930), Prof. Dingle makes a key statement which is completely clear, free of ambiguity—and incorrect. This statement, and this alone, leads him from his preceding correct results directly to his incorrect conclusion that no asymmetrical ageing occurs in the familiar round trip. I quote Prof. Dingle. (I have enclosed the crucial incorrect statement and resulting incorrect conclusion in brackets { }, and have added my own Fig. 1 to illustrate Prof. Dingle's notation.)

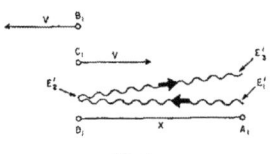

Fig. 1

"Consider three clocks, B_1, C_1, D_1, which, at the common reading $t = 0$, are together and such that, if D_1 is regarded as being at rest, B_1 and C_1 are moving in opposite directions, each with uniform speed V with respect to D_1. If these clocks are respectively at the origins of co-ordinate systems of which the x-axes all lie along the line of motion and are positive in the direction towards which C_1 is moving, then these systems are related by the Lorentz transformation. There is a fourth clock A_1 at rest at the point $(x,0,0)$ in the D_1 system, and at the event $E_1' = (x,0,0, -x/c)$ in that system let it emit a beam of light which, on reaching D_1, is reflected back again to A_1, clearly reaching it at the event $E_2' = (x,0,0,x/c)$ in the D_1 system. By the Lorentz transformation, $E_1' = (x\beta,0,0, -x\beta/c)$ in the B_1 system and $(x/\beta,0,0, -x/c\beta)$ in the C_1 system, where $\beta = [(1 - V/c)/(1 + V/c)]^{1/2}$, while $E_2' = (x/\beta,0,0,x/c\beta)$ in the B_1 system and $(x\beta,0,0,x\beta/c)$ in the C_1 system. Hence, by Einstein's criterion, the time of the event E_2', at which the light falls on D_1, will be 0 in the D_1 system, { $(1/2(-x\beta/c + x/c\beta)$ in the B_1 system,

and $1/2(-x/c\beta + x\beta/c)$ in the C_1 system)}. Since D_1 is at rest with respect to A_1, D_1 is synchronized with A_1 if it actually reads 0 at this event. Let this be so. Then, by hypothesis, B_1, C_1 and D_1 will be coincident and will all read 0 at the event E_2'. {(Hence B_1 will be slow and C_1 fast)} by the same amount, viz., $x(1/\beta - \beta)/2c$, at the event E_2'."

If Prof. Dingle had used again the same two Lorentz transformations that gave him his (correct) values for the times of events E_1' and E_2' (emission and absorption) as measured in the rest systems of B_1 and C_1, he would have found that the reflexion E_2' occurs at zero time in those systems as well as in the D_1 system; therefore clocks B_1 and C_1 are neither slow nor fast, but are both 'on time'.

Prof. Dingle's criterion, the use of the expression $t_2 = \frac{1}{2}(t_1 + t_3)$ to synchronize a local stationary clock with a distant *moving* clock, is incompatible with the Lorentz transformation, and has never before been used, either in Einstein's first paper[7], or in any subsequent conventional presentation[8]. Einstein used the above expression only to define "synchronous stationary clocks located at different places"[7]. To synchronize clocks in two co-ordinate systems in uniform relative motion he used an entirely different procedure—the setting to zero of just one clock in each system at the instant when these two particular clocks pass one another and are at the same place. (These clocks are usually taken to be at the origin of co-ordinates in each system.) These two types of synchronizations, and the invariance of c, determine the Lorentz transformation completely. There is no more freedom to satisfy Prof. Dingle's criterion. Straightforward application of the Lorentz transformation to Prof. Dingle's example shows that in B_1's rest frame the reflexion E_2' occurs in general at $t_2 = \frac{1}{2}(t_2 + t_1) - \frac{1}{2}(V/c)(t_3 - t_1)$. Prof. Dingle, by using his incorrect criterion instead of the Lorentz transformation, omitted the term in V. Insertion of Prof. Dingle's (correct) values for t_1 and t_3 leads to $t_2 = 0$. Similarly, in the C_1 frame, application of the Lorentz transformation shows that E_2' occurs at zero time. The reflexion therefore occurs at zero time in all three frames.

The remainder of Prof. Dingle's derivation is correct, and shows, after correcting the above error, the usual result—that the special theory of relativity predicts that the asymmetrical age difference acquired in the round trip is twice that acquired in the outgoing one-way trip (if the return trip is taken at the same velocity).

In summary, Prof. Dingle's regrettable error is completely accounted for by his failure to notice Einstein's clear and explicit statement in section I of his famous paper[7], that the synchronization there defined applies only to clocks having no relative motion.

We must now consider Prof. Dingle's qualitative argument, since it is at first sight rather convincing. Furthermore, Prof. Dingle has himself said that if some error were found in his quantitative argument (as I have done), then he would still believe in the qualitative argument.

Prof. Dingle argues that we cannot distinguish by any experiment which brother remains 'at rest', during the round trip, because, he says, according to the principle of relativity only the relative motion of the brothers has any physical meaning. It is therefore, he says, a matter of arbitrary convention which brother we regard as being at rest. It follows

Reprinted from Nature
(May 1957 - Pages 1071-1072)
By Permission

that any observed asymmetry would lead to a contradiction with the principle of relativity, since we could use the asymmetry to determine which brother remained at 'absolute rest'.

This argument is wrong. It holds only for continued uniform relative motion. For general motion, including relative accelerations (which are, of course, necessary if the brothers are to start and end with no separation in space, and no relative velocity), one must consider also the (relative) motion of the brothers with respect to an important third body—the remaining matter in the universe. This was first demonstrated experimentally by Newton with his famous water-pail experiment. Newton showed that for accelerated motion (here circular), the Earth, and every frame in uniform motion with respect to the Earth, is a 'preferred frame', that is, one in which Newton's laws are valid. For linear accelerations the same is obviously true, as can most easily be demonstrated by a modified experiment wherein a pail of water is connected by a spring to an equal mass of, say, lead. The whole system is placed on a frictionless surface and set into vibration. The behaviour of the liquid surface is completely different, depending on whether it is the lead weight or the pail of water that we then hold fixed with respect to the Earth, although the relative velocities and accelerations of the pail and lead are in either case the same. This experiment completely demonstrates the falsity of Prof. Dingle's argument that only the relative motion of the brothers matters. Presumably in an otherwise empty universe Prof. Dingle's argument would hold.

Newton's conclusion from the water-pail experiment was that the Earth was in uniform motion with respect to 'absolute space'. Mach, and later Einstein[9], substituted 'the distant galaxies' for Newton's unsatisfactory 'absolute space'. By thus considering the 'third body'—the remainder of the universe—the theory of relativity was extended to encompass accelerated relative motion of co-ordinate systems, so as to eliminate the unsatisfactory favoured role taken in the special theory of relativity by the 'inertial frame'. Once that role has been so clarified, however, we are again free to calculate with the concepts of special relativity, even in problems involving relative accelerations.

In the clock paradox trip, the source of asymmetry is not the relative acceleration of the two twins; rather, it is the relative acceleration of each twin separately with respect to the 'third body'—the inertial frames (all frames in uniform motion with respect to the distant galaxies). The 'stay at home' twin remains in uniform motion with respect to the universe, not to 'absolute space', and the principle of relativity is satisfied. (It does not matter whether the travelling twin is regarded as accelerated with respect to the universe and the stay-at-home twin, or whether, instead, the universe and stay-at-home twin are accelerated.)

In working out the problem by the use of the concepts of special relativity alone, the trip can be chosen in such a manner that the duration of the acceleration periods has a negligible effect on the relative asymmetrical ageing, so that the accelerations need not be considered explicitly in working the problem[3]. Nevertheless, the accelerations completely determine the problem, through the crucial (though somewhat hidden) role of the inertial frame as "a frame in which special relativity holds"[4,10].

I would like to express my appreciation to Prof. Edwin M. McMillan for a number of useful discussions.

FRANK S. CRAWFORD, JUN.

Radiation Laboratory,
University of California,
Berkeley 4.

[1] Einstein, A., Ann. Physik., 17, 891 (1905). See sec. 4. All the publications of Albert Einstein to which I refer in this article are found in English translation in the recently republished book by H. Lorentz, A. Einstein, H. Minkowski and H. Weyl, "The Principle of Relativity" (Dover Publications, Inc.).

[2] Dingle, H., Nature, 177, 782 (1956).

[3] McCrea, W. H., Nature, 167, 680 (1951).

[4] Einstein, A., Ann. Physik, 17, 891 (1905), see sec. 1, first sentence.

[5] Dingle, H.; McCrea, W. H., Nature, 178, 680 681 (1956). Dingle, H., Nature, 179, 865 (1957).

[6] Dingle, H., Proc. Phys. Soc., A, 69, 925 (1956).

[7] Einstein, A., Ann. Physik, 17, 891 (1905), see last half of sec. 1.

[8] Møller, C., "The Theory of Relativity" (Oxf. Univ. Press, London, 1952).

[9] Einstein, A., Ann. Physik, 49, 769 (1916), see sec. 2.

[10] Einstein, A., Ann. Physik, 49, 769 (1916), see sec. 1, first paragraph.

LETTERS TO THE EDITORS

The Editors do not hold themselves responsible for opinions expressed by their correspondents. No notice is taken of anonymous communications.

The Clock Paradox in Relativity

IN the course of reasoning on this subject with some of my more recalcitrant friends, I have come across a numerical example which I think makes the matter easier to follow than would any mathematical formulæ, and perhaps this might interest some readers of *Nature*.

There is no doubt whatever that the accepted theory of relativity is a complete and self-consistent theory (at any rate up to a range of knowledge far beyond the present matter), and it quite definitely implies that a space-traveller will return from his journey younger than his stay-at-home twin brother. We all of us have an instinctive resistance against this idea, but it has got to be accepted as an essential part of the theory. If Prof. H. Dingle should be correct in his disagreement, it would destroy the whole of relativity theory as it stands at present.

Some have found a further difficulty in understanding the matter. When two bodies are moving away from each other, each sees the occurrences on the other slowed down according to the Doppler effect, and relativity requires that they should both appear to be slowed down to exactly the same degree. Thus if there are clock-dials on each body visible from the other, both will appear to be losing time at the same rate. Conversely, the clocks will appear to be gaining equally as they approach one another again. At first sight this might seem to suggest that there is an exact symmetry between the two bodies, so that the clock of neither ought in the end to record a time behind that of the other. The present example will show how this argument fails.

In order to see how a time-difference will arise, it suffices to take the case of special relativity without complications from gravitation. Two space-ships, S_0 and S_1, are floating together in free space. By firing a rocket S_1 goes off to a distant star, and on arrival there he fires a stronger rocket so as to reverse his motion, and finally by means of a third rocket he checks his speed so as to come to rest alongside S_0, who has stayed quietly at home all the time. Then they compare their experiences. The reunion of the two ships is an essential of the proceedings, because it is only through it that the well-known difficulties about time-in-other-places are avoided.

The work is to be so arranged that it can be done by ordinary ships' navigators, and does not require the presence in the crews of anyone cognizant of the mysteries of time-in-other-places. To achieve this, I suppose that the two ships are equipped with identical cæsium clocks, which are geared so as to strike the hours. On the stroke of every hour each ship sends out a flash of light. These flashes are seen by the other ship and counted, and they are logged against the hour strokes of its own clock. Finally the two logs will be compared.

In the first place it must be noted that S_1's clock may behave irregularly during the short times of his three accelerations. This trouble can be avoided by instructing him to switch the clock off before firing his rockets, and only to start it again when he has got up to a uniform speed, which he can recognize from the fact that he will no longer be pressed against one wall of his ship. The total of his time will be affected by this error, but it will be to the same extent whether he is going to the Andromeda Nebula, or merely to Mars. Since the time that is the subject under dispute is proportional to the total time of his absence, this direct effect of acceleration can be disregarded.

I choose as the velocity of S_1's travel $v = \frac{4}{5}c$, because in this special case there are no tiresome irrationalities to consider. I take the star to be 4 light-years away from S_0. The journey there and back will therefore take 10 years according to S_0. Immediately after the start each will observe the other's flashes slowed down by the Doppler effect. The formula for this in relativity theory is $\sqrt{(c+v)/(c-v)}$, which in the present case gives exactly 3. That is to say, each navigator will log the other's flashes at a rate of one every three hours of his own clock's time. Conversely, when they are nearing one another again, each will log the other's flashes at a rate of three an hour.

So far everything is perfectly symmetrical between the ships, but the question arises, for each ship respectively, how soon the slow flashes will change over into fast ones. First take the case of S_1. During his outward journey he will get slow flashes, but when he reverses direction at the star, they will suddenly change to fast ones. Whatever his clock shows at this time it is certainly just half what it will show when he gets home. Thus for half the journey he will get flashes at a rate of $\frac{1}{3}$ per hour, and for the other half at a rate of 3 per hour. The average for the whole journey will thus be at a rate $\frac{1}{2}\left(\frac{1}{3}+3\right) = \frac{5}{3}$ per hour. During this time S_0 will have sent out 10 years' worth of flashes, and so in the end S_1's clock will record $\frac{3}{5} \times 10 = 6$ years, which, of course, he can verify directly from his detailed log.

S_0's log will be quite different. He will start with slow flashes and end with fast ones, but the change-over is determined by S_1's reversal, which is occurring 4 light-years away from him. Consequently, he will get slow flashes for $5 + 4 = 9$ years, and therefore fast flashes for only 1 year. The total number he will count is $\frac{1}{3} \times 9 + 3 \times 1 = 6$ years' worth. His nine years of slow flashes and one of fast are in marked contrast with S_1's experience of three years of each. Thus when the navigators compare their logs together they will be completely different, but both will agree that S_0's clock went for ten years and S_1's for only six.

It may be that S_0 will suggest that for some reason S_1's clock was going slow during the motion, but S_1 will point out that there was no sign of anything wrong with it, and that anyhow his heart-beat and other bodily functions matched the rate of his clock, and he may even direct attention to the fact that his forehead is perceptibly less wrinkled than that of his twin brother. In fact—as the relativist knows—he is now actually four years younger than his brother.

In giving this example, I have assumed S_0 at rest for the sake of simplicity, but it is not hard to verify that the two logs will be exactly the same if a uniform motion of any amount is superposed on the system. However, to show this would go beyond the scope of this communication.

C. G. DARWIN

Newnham Grange,
  Cambridge.
  Sept. 30.

Reprinted from Nature
(Nov. 1957 - Pages 976-977)
By Permission

Relativity and Space Travel*

J. R. PIERCE†, FELLOW, IRE

Summary—This paper treats in terms of the special theory of relativity a "clock paradox" involving the fact that the frequency of an atomic oscillator on a moving body is lowered but the mass which is converted into radiation is increased; the case of the twin who goes on a space trip at near-light speed and returns younger later than his brother on earth; the shift in frequency in the presence of a gravitational field; the clock rate on a satellite; the speed attainable by a photon rocket; and a space ship propelled by the energy of interstellar matter.

I. INTRODUCTION

TODAY engineers are building rockets and satellites with velocities unprecedented among gross, man-made objects. Further, they are seriously speculating concerning even higher velocities in connection with interplanetary travel. There have been repeated suggestions that a "photon rocket" could be used to attain a speed very close to that of light.

It is well known that the theory of Newtonian mechanics is inaccurate in dealing with velocities appreciable compared with the velocity of light. The solution to the problem of applying relativity consistently to the practical and speculative problems of space flight is less well known, and various questions and errors recur repeatedly.

Fortunately, it is possible to treat many interesting problems of space travel by means of special relativity, together with a few reasonable *ad hoc* assumptions. This is the purpose of the present paper. Portions of the material will be found in a number of sources.[1]-[5] A reasonably complete and unified account is attempted here.

Before we proceed to particular problems, a few general remarks may be helpful. In making use of relativity, it is important to avoid certain easy pitfalls. One of these is the notion of simultaneity. We make matters more difficult for ourselves, for instance, if in reckoning time on a space ship we talk about the time *on earth* at which the space ship turns around and heads toward earth instead of away from it. This can lead to a seeming "clock paradox" which will be discussed later. Feynman has expounded the matter of simultaneity in a wonderfully simple manner. In pursuing his explanation we should note that, according to special relativity, an object going past us appears to be foreshortened in the direction of motion.

Imagine two very long starships, ships which to us are equally long, passing each other while traveling in opposite directions, as shown in Fig. 1. Suppose that just as they are opposite one another, two bolts of lightning pass instantly between them, one from the nose of the upper, leftward-traveling ship, which we will call ship *L*, to the tail of the lower, rightward-traveling ship, ship *R*, and the other from the tail of ship *R* to the nose of ship *L*. An observer in the center of the upper ship, ship *L*, will see the left-hand bolt first, because he tends to catch up with the light emitted by it, while he travels away from the light emitted by the bolt to the right. On the other hand, an observer in the center of ship *R* will see the light from the right-hand bolt first, for he rushes to meet the light from the right and flees the light from the left.

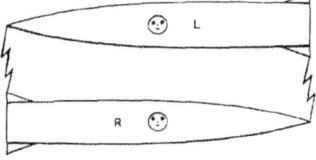

Fig. 1.

Special relativity tells us that the man in ship *L* has a perfect right to say, "I am standing still, and if I see the left-hand bolt of lightning, the one at my nose, before the right-hand bolt, the one at my tail, it is because the bolt at my nose occurred first. Indeed, it had to, for I see that that other ship, ship *R*, is much shorter than my ship, ship *L*, and hence his tail was opposite my nose before his nose reached the position of my tail."

On the other hand, the man in ship *R* says, "I am standing still. I saw the right-hand bolt, the one at my nose, first, and so it occurred first. Indeed, it had to, for that ship *L* is much shorter than my ship *R*, and his tail was opposite my nose before his nose reached the position of my tail."

Clearly, no argument can resolve this contradiction as to which bolt of lightning struck first, if, indeed, either did. An assumption of some instantaneous form of communication (telepathy, for instance) which enables us to determine a "true" time order of events would involve us in a hopeless paradox. We simply can't have such a thing and relativity too: one or the other must go.

* Original manuscript received by the IRE, April 1, 1959.
† Bell Telephone Labs., Inc., Murray Hill, N. J.
[1] J. H. Jeans and E. Whittaker, "Relativity," in "The Encyclopedia Brittanica," vol. 19, pp. 89–98, 1955.
[2] J. J. Coupling, "On atomic jets," *Astounding Science Fiction*, vol. 54, pp. 115–127; January, 1955.
[3] S. F. Singer, "Application of an artificial satellite to the measurement of the general relativity 'red shift'," *Phys. Rev.*, vol. 104, pp. 11–14; October 1, 1956.
[4] R. Schlegel, "New clock problems in special relativity," *Bull. Am. Phys. Soc.*, vol. 2, p. 239; April 25, 1957.
[5] C. G. Darwin, "The clock paradox in relativity," *Nature*, vol. 180, pp. 976–977; November 9, 1957.

Reprinted from Proc. of the IRE
(June 1959 - Pages 1053-1061)
By Permission

As the relative velocities of the two ships approach the speed of light, one captain will insist that event A occurs a time L/c before event B, while the other insists that event A occurs at time L/c after event B, there will be a possible disagreement of $2L/c$ about what is simultaneous with what, where L is the length that each assigns to his ship. In general, for events occurring a distance L apart, there is a range of time $2L/c$ over which we cannot say whether one event occurred before or after another.

We can assert that event B at location b occurred after event A at location a, which is a distance L away only when a light signal generated by event A at a reaches b before event B occurs; this takes a time L/c. The same statement holds if we interchange A and a with B and b.

II. Some Equations of Special Relativity

It is not the purpose of this paper to derive the equations of special relativity. The purpose is to explain them sufficiently so that they can be used correctly and to apply them to certain problems in space travel.

The equations we need deal with an observer who considers himself to be standing still and a space ship, planet, solar system or galaxy which is moving past him, from left to right, with a velocity v. As we noted earlier, to the observer, all objects in the moving system appear to be shrunk or contracted by a factor which we will call a shrinking factor S. This factor is

$$S = \sqrt{1 - (v/c)^2}. \quad (1)$$

Here c is the velocity of light.

$$c = 3 \times 10^8 \text{ meters/sec.} \quad (2)$$

The moving object appears shrunk only in the direction of motion, not crosswise to it.

The clocks on the moving object also appear to be going slower by this same shrinking factor S, so that when our clock indicates the passage of one hour, the clocks on the moving system indicate the passage of only a fraction S of an hour.

Let us now consider the people in the moving system. One of them shoots a projectile in the direction of motion and says, "I have shot this forward with a velocity u_m." But, when we measure the velocity by our "fixed" standards, we obtain a velocity u

$$u = \frac{u_m + v}{1 + vu_m/c^2}. \quad (3)$$

We note that the velocity u which we observe can never be greater than c, the velocity of light. Suppose, for instance, that a man in the moving system shines a beam of light to the right, a beam which he says has the speed of light, so that $u_m = c$. Both he and the beam are rushing to the right, he with a velocity v. Yet, if we put $u_m = c$ into (3), we get for u simply c, the velocity of light. Eq. (3) will not be used in the subsequent work, but it is an important tool.

When we observe matter in motion with a velocity v, it appears to have more mass. Thus, something which to a man in the moving system appears to have a mass m_0 (called the rest mass) appears to us, past whom it is moving with a velocity v, to have a mass m given by

$$m = \frac{m_0}{\sqrt{1 - (v/c)^2}}. \quad (4)$$

The rest mass is the quantity of matter; it is what remains constant when we accelerate a body, while the relativistic mass m increases.

We are all familiar with the relativistic expression for energy E,

$$E = mc^2. \quad (5)$$

This is universally applicable. As a moving object has more mass than it would if it were standing still [according to (4)], (5) tells us that it will also have more energy. The additional energy is of course the kinetic energy of the body, that is, the energy associated with its motion.

In order to make calculations concerning space travel, we must associate with these laws of relativity two universal and basic laws of physics: the conservation of energy and the conservation of momentum. These laws say that the total energy and momentum must be the same before and after a physical event, such as the acceleration of a starship.

We can express the energy of the system in terms of its masses, at rest or moving, by means of relation (5). As the energy of each mass is proportional to the mass by the same constant, c^2, we see that the conservation of energy means the conservation of mass. Rest mass may be diminished, but if we use the energy produced to set matter in motion, the relativistic masses [as given by (4)] total the same as do the original masses. Or, electromagnetic radiation such as light may be produced, which, according to relativity, also has mass.

The momentum p of a material body of mass m and velocity v is

$$p = mv = \frac{m_0 v}{\sqrt{1 - (v/c)^2}}. \quad (6)$$

Momentum of course has a direction; this is the same direction as the velocity of the body.

According to relativity, electromagnetic radiation, including gamma rays, X rays, light and radio waves, must have mass and momentum as well as energy. Let us call the energy of a certain amount of radiation E_r, its mass m_r and its momentum p_r. The radiation travels with the velocity of light, c. The energy, mass and momentum are related by the following equations:

$$E_r = m_r c^2 \quad (7)$$
$$p_r = E_r/c = m_r c. \quad (8)$$

We see that (7) and (8) are really just the same as (5) and (6). However, radiation has no rest mass m_0.

We are now equipped with all the physical laws we need for our calculations, except a quantum law governing radiation, which we will encounter in the next section.

III. A "CLOCK PARADOX"

In accord with the conservation of energy and the fact that energy is mc^2, when radiant energy such as light appears, matter must disappear. Thus, when an atom or molecule emits a quantum of light or other electromagnetic radiation, such as radio waves or gamma rays, it loses energy. Quantum mechanics tells us something more about this process. The energy of a quantum of light or other radiation is Planck's constant h times the frequency of the light, f. Hence, from the conservation of energy,

$$hf = m_r c^2 = E_r \qquad (9)$$

or

$$f = m_r c^2/h = E_r/h \qquad (10)$$

$$h = 6.55 \times 10^{-34} \text{ joule/sec.} \qquad (11)$$

Here m_r is the mass of the quantum of radiation and it is also equal to the mass lost by the atom or molecule in producing the radiation. E_r is the energy of the radiation.

Relativity tells us that the mass of each atom or molecule *increases* when the atom or molecule is traveling fast. Further, our best clocks are regulated by the frequencies of radiation of ammonia molecules or cesium atoms. If we took such a clock on a swiftly-moving space ship, the mass of each molecule or atom would increase, and hence the small fraction of matter which it would lose in emitting radiation would also have more mass. Thus, we might expect that the frequency of the electromagnetic wave produced by the atom or molecule would be higher than if the ship were standing still. Since this frequency governs the speed of our clocks, we might expect the clocks on the ship to go faster by a factor $1/\sqrt{1-(v/c)^2}$.

Relativity tells us, quite to the contrary, that the rate of the moving clocks is decreased by a factor

$$S = \sqrt{1 - (v/c)^2}. \qquad (1)$$

How are we to explain this seeming paradox?

Let us explore this matter warily. As a first step, let us assume that we send waves of light or other radiation from earth to a moving space ship. What frequency will these waves seem to have to an observer on the space ship?

We will consider two cases, shown in Fig. 2. In (a), the ship is headed toward the waves of the radiation with a velocity v, and of course the waves of radiation go toward the ship with the velocity c. If λ is the wavelength of the waves, that is, the distance between wave crests, then the frequency f of the waves, that is, the number of crests λ apart which will strike a fixed object in a second, is clearly

$$f = c/\lambda. \qquad (12)$$

However, we see the ship and the waves as having a relative velocity $c+v$, so we see the wave crests as striking the ship with a frequency f_1 given by

$$f_1 = \frac{c + v}{\lambda} = (1 + v/c)f. \qquad (13)$$

Is this the frequency at which the man on the ship observes the wave crests to arrive? No, for his clock runs slow by a factor $\sqrt{1-(v/c)^2}$. Thus, to him the crests seem to arrive more rapidly, with a frequency f_d given by

$$f_d = \frac{f_1}{\sqrt{1 - (v/c)^2}}. \qquad (14)$$

By using the value of f_1 given by (13) in (14) we find

$$f_d = \frac{\sqrt{1 + v/c}}{\sqrt{1 - v/c}} f. \qquad (15)$$

This is known as the *relativistic Doppler frequency*.

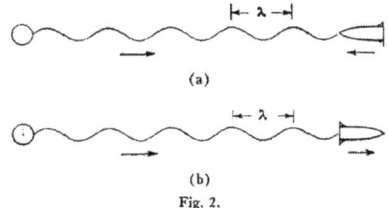

(a)

(b)

Fig. 2.

In the case shown in Fig. 2(b), in which the ship is going away from the source of radiation, we see the waves of radiation overtaking the ship with a relative velocity

$$c - v.$$

Accordingly, the frequency f with which *we* see crests strike the ship is

$$f_1 = (1 - v/c)f. \qquad (16)$$

However, because the ship's clock runs slow, those on the ship observe a frequency

$$f_d = \frac{\sqrt{1 - v/c}}{\sqrt{1 + v/c}} f. \qquad (17)$$

So far, we have stood still on earth and watched an observer on the moving ship measure the frequency of the radiation that we send him. But who is to say which is moving, the ship or the earth? Suppose that the captain of the ship watches an atom on the (to him) rapidly-moving earth emit radiation. The energy produced must be the change in relativistic mass, which is greater than the rest mass, times the square of the velocity of light. Since the speed of the earth results in an increase in the mass, does this increase in mass directly account for the frequency of radiation emitted? No, it does not!

What we do know is that both energy and momentum must be conserved in the emission of radiation from the sources on earth. Let us assume that the radiation is emitted downward, at an angle θ with respect to the normal to the path of the earth, which travels to the right with a velocity v, as shown in Fig. 3.

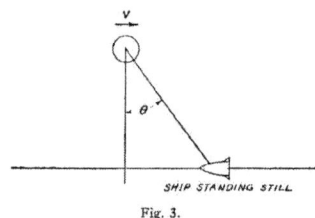

SHIP STANDING STILL

Fig. 3.

Let us consider the energy and momentum of an object traveling to the right with a velocity v which has x and y components v_x and v_y. The magnitude v of the velocity is

$$v = \sqrt{v_x^2 + v_y^2}. \tag{18}$$

The energy E will be

$$E = \frac{m_0 c^2}{\sqrt{1 - (v_x/c)^2 - (v_y/c)^2}}. \tag{19}$$

The x and y components of momentum, p_x and p_y, will be

$$p_x = \frac{m_0 v_x}{\sqrt{1 - (v_x/c)^2 - (v_y/c)^2}} \tag{20}$$

$$p_y = \frac{m_0 v_y}{\sqrt{1 - (v_x/c)^2 - (v_y/c)^2}}. \tag{21}$$

Let us suppose that the rest mass m_0 is changed by a small amount dm_0 and that the x and y components of velocity are changed by small amounts dv_x and dv_y. By differentiation, we find the changes in energy and momenta to be

$$dE = \frac{m_0(v_x dv_x + v_y dv_y)}{(1 - (v_x/c)^2 - (v_y/c)^2)^{3/2}}$$
$$+ \frac{c^2 dm_0}{\sqrt{1 - (v_x/c)^2 - (v_y/c)^2}} \tag{22}$$

$$dp_x = \frac{m_0(1 - (v_y/c)^2)dv_x}{(1 - (v_x/c)^2 - (v_y/c)^2)^{3/2}}$$
$$+ \frac{v_x dm_0}{\sqrt{1 - (v_x/c)^2 - (v_y/c)^2}} \tag{23}$$

$$dp_y = \frac{m_0(1 - (v_x/c)^2)dv_y}{(1 - (v_x/c)^2 - (v_y/c)^2)^{3/2}}$$
$$+ \frac{v_x dm_0}{\sqrt{1 - (v_x/c)^2 - (v_y/c)^2}}. \tag{24}$$

Let us now apply (22)–(24) to our case, in which the earth is whizzing by in the x direction with a velocity v. In this case

$$v_y = 0$$
$$v_x = v. \tag{25}$$

Accordingly,

$$dE = \frac{m_0 v dv_x}{(1 - (v/c)^2)^{3/2}} + \frac{c^2 dm_0}{\sqrt{1 - (v/c)^2}} \tag{26}$$

$$dp_x = \frac{m_0 dv_x}{(1 - (v/c)^2)^{3/2}} + \frac{v dm_0}{\sqrt{1 - (v/c)^2}} \tag{27}$$

$$dp_y = \frac{m_0 dv_y}{\sqrt{1 - (v/c)^2}}. \tag{28}$$

Now let E_r be the energy of the emitted radiation. The earth, in emitting the radiation, must lose an energy equal to the energy of the radiation, so that we must have

$$dE = -E_r. \tag{29}$$

Further, in order for momentum to be conserved, we must have

$$dp_x = -(E_r/c) \sin \theta \tag{30}$$
$$dp_y = (E_r/c) \cos \theta. \tag{31}$$

Let us further call the rest mass which a molecule or atom on earth loses in emitting radiation m_r. Thus

$$dm_0 = -m_r. \tag{32}$$

By substituting (29)–(32) into (26)–(28), we obtain

$$E_r = \frac{-m_0 v dv_x}{(1 - (v/c)^2)^{3/2}} + \frac{m_r c^2}{\sqrt{1 - (v/c)^2}} \tag{33}$$

$$(E_r/c) \sin \theta = \frac{-m_0 dv_x}{(1 - (v/c)^2)^{3/2}} + \frac{m_r v}{\sqrt{1 - (v/c)^2}} \tag{34}$$

$$(E_r/c) \cos \theta = \frac{m_0 dv_y}{\sqrt{1 - (v/c)^2}}. \tag{35}$$

We note that dv_y appears in (35) only. This equation can be regarded as giving dv_y directly in terms of E_r. We can eliminate dv_x by using (33) and (34); we obtain

$$E_r = \frac{\sqrt{1 - (v/c)^2}}{1 - (v/c) \sin \theta} m_r c^2. \tag{36}$$

Here $m_r c^2$ is the energy of a photon as seen from earth, while E_r is the energy of the photon according to an observer who says that the earth is moving past him with a velocity v, as shown in Fig. 3. As frequency is proportional to energy, then if f is the frequency of a photon as observed on earth and f_d as observed by our "fixed" observer, (36) tells us that

$$f_d = \frac{\sqrt{1 - (v/c)^2}}{1 - (v/c)\sin\theta} f. \qquad (37)$$

When $\theta = \pi/2$, so that $\sin\theta = 1$, we have the case of earth and ship approaching, to which (15) applies, and we are gratified to see that (37) indeed reduces to (15). When $\theta = -\pi/2$, so that $\sin\theta = -1$, we have the case of earth and ship separating, and indeed, (37) reduces to (17), which applies to this case.

When $\theta = 0$, so that $\sin\theta = 0$, the earth is neither approaching nor receding from the ship. Here we observe the oscillator frequency of the ship, unaffected by relative motion; we see from (37) that it is

$$\sqrt{1 - (v/c)^2} f. \qquad (38)$$

This is exactly in accord with the shrinking factor S of (1).

At first thought, it may seem puzzling that while the relativistic mass of the matter lost during radiation of the quantum is $m_r/\sqrt{1 - (v/c)^2}$, the mass of the radiation produced is only $m_r\sqrt{1 - (v/c)^2}$. We note, however, that for momentum to be conserved in the x direction after the earth has lost rest mass, the earth must travel faster, and this means that the rest of the earth must gain energy during the process of radiation.

We may also note that radiation which to the fixed observer appears to be traveling normal to the x axis must appear to an observer on the "moving" earth to have a component of motion in the $-x$ direction. In illustration, we may note that a swiftly flying airplane would have to eject a bomb violently backward in order for it to fall straight down to a man on earth.

IV. THE CLOCK PARADOX

We have just considered *a* clock paradox. This is not, however, *the* clock paradox that most people mean when they use the words. In the clock paradox, one of two identical twins goes off in a starship at 99.5 per cent of the speed of light. As observed from earth, his clock runs only 1/10 normal speed. After five years pass on earth, the ship turns around, and 10 years from takeoff he lands on earth. The twin steps out, but he has aged only 1 year, while his brother who stayed on earth is 10 years older than when the ship started.

However, it has been argued that to the twin on the ship, the clock on earth should have appeared to be running at 1/10 normal speed. Should we say that the twin on earth ages only one year while the twin on the ship ages 10 years?

Such arguments are confusing because they involve one in considerations of the time on distant earth when the ship turns around, or the time on the ship as reckoned from earth when the ship turns around. As we have seen, what events (perhaps clock readings) we take to be simultaneous depend on velocity.

We can avoid a lot of trouble if we make only such time comparisons as involve objects at the same point. Among such objects we may include electromagnetic waves emitted by the ship or by the earth as well as material objects.

How do we measure the clock rate on some object in motion with respect to ourselves, anyway? We can infer this rate from the relativistic Doppler frequency. Let us, however, work with this approach directly and see what sort of answers the approach leads us to.

First, let us consider the starship from the point of view of earth people. As indicated in Fig. 4, the ship goes out at a velocity v toward some point * (perhaps a star). It then turns around and returns at the same velocity v. The point * is at a distance L from earth as measured by earthmen.

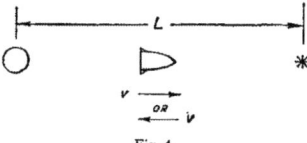

Fig. 4.

Suppose that the ship carries a standard oscillator just like one left on earth. Each oscillator would oscillate with the same frequency f if it were standing still. Now, let us measure elasped time in terms of cycles of oscillation of these oscillators. On earth, the time t_e in seconds between the departure and the arrival of the starship will be

$$t_e = \frac{2L}{v}. \qquad (39)$$

The number of cycles of oscillation N_e of the earth clock is thus

$$N_e = ft_e = \frac{2Lf}{v}. \qquad (40)$$

What about radiation received from the ship? On the outward journey, when the ship is going away from earth, the Doppler frequency f_{d1} received at earth will be less than f; in fact, (17) says that it will be

$$f_{d1} = \frac{\sqrt{1 - v/c}}{\sqrt{1 + v/c}} f. \qquad (41)$$

But, radiation of this frequency is emitted from the ship right up to the time it turns around, when it is a distance L from the earth. The last of this radiation, which travels toward earth with a velocity c, will reach earth at a time L/c later than the time at which the ship turns around. Hence, the radiation received on earth has the frequency f_{d1} for a time

$$L/v + L/c \qquad (42)$$

out of the total time of the journey, $2L/v$. For the rest of the time, which must be

$$2L/v - (L/v + L/c) = L/v - L/c, \qquad (43)$$

the Doppler frequency received will be that for a source approaching the earth, as given by (15); we will call this frequency f_{d2}.

$$f_{d2} = \frac{\sqrt{1 + v/c}}{\sqrt{1 - v/c}} f. \qquad (44)$$

To get the total number of cycles of radiation N_s which have been produced on the ship and have struck the earth between the time the ship left and the time it returned, we should multiply the time (42) by the frequency (41) and add the product of the time (43) and the frequency (44). Thus

$$N_s = \frac{Lf}{v}\left[(1 + v/c)\frac{\sqrt{1 - v/c}}{\sqrt{1 + v/c}} + (1 - v/c)\frac{\sqrt{1 + v/c}}{\sqrt{1 - v/c}} \right]$$

$$N_s = \frac{2Lf}{v}\sqrt{1 - (v/c)^2}. \qquad (45)$$

We see by comparing (40) and (45) that according to this calculation, the oscillator on the ship must have run slower than the oscillator on earth by just the proper relativistic time shrinking factor, S of (1).

But, suppose that we are on the ship? Here the picture is as shown in Fig. 5. The universe is whizzing by in one direction or the other with a velocity v. Because of this, the distance between earth and * has shrunk relativistically to a distance

$$L\sqrt{1 - (v/c)^2}. \qquad (46)$$

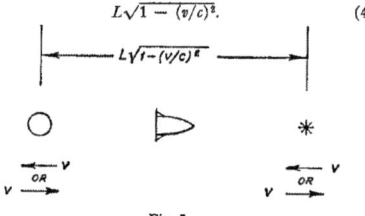

Fig. 5.

Accordingly, on the ship the "trip" from earth to * and back takes a time t_s

$$t_s = \frac{2L\sqrt{1 - (v/c)^2}}{v}. \qquad (47)$$

The number of cycles N_s emitted by the ship's oscillator in this time is

$$N_s = ft_s = \frac{2Lf\sqrt{1 - (v/c)^2}}{v}. \qquad (48)$$

What about the radiation from earth? On the "journey out" the earth will appear to be moving away from us with a velocity v for a time half that of (47), which is the time for the two-way trip. But, when we reach * we change our motion with respect not only to the earth and the star, but with respect to the radiation from the earth as well. The waves of radiation which we immediately encounter are already in space; we rush to meet them, and the observed frequency immediately changes from that of (17) to that of (15). Hence, we have the lower and the higher Doppler frequencies each for the same time. Thus, the total number of cycles N_e which leave earth and fall on the ship during both parts of the trip is

$$N_e = \frac{Lf\sqrt{1 - (v/c)^2}}{v}\left[\frac{\sqrt{1 - v/c}}{\sqrt{1 + v/c}} + \frac{\sqrt{1 + v/c}}{\sqrt{1 - v/c}} \right]$$

$$N_e = \frac{2Lf}{v}. \qquad (49)$$

Let us now compare the calculations made by the earthbound observer with those made by the observer on the ship. The results are exactly the same. N_s from (45) is exactly N_s from (48), and N_e from (40) is exactly N_e from (49). If we work matters out carefully, there is no clock paradox involved. The crew on the starship age less rapidly than those left behind on earth.

It may of course be objected that the acceleration of turning does dire things to the ship's clock. Here we can at least say that the effect should be the same for a long as for a short trip, and could only be additive.

The reader can easily arrive at the above result in another simple way. Assume two ships, the first going from the earth to * with a velocity v and the second going from * to the earth with a velocity v. Assume that the ships pass one another at *. Assume that ship 1 and earth set their clocks to zero as ship 1 passes earth. Imagine that ship 2 sets its clock by the clock of ship 1 as ship 2 passes ship 1. Let earth and ship 2 compare clocks as ship 2 passes earth. It will be found that the reading of the clock of ship 2 (which represents elasped time of ship 1 from earth to star plus elapsed time of ship 2 from star to earth) is smaller than the reading of the earth clock by the factor $\sqrt{1 - (v/c)^2}$, no matter what frame of reference we use.

V. Frequencies on a Ship with a Changing Velocity

Consider the moving space ship shown in Fig. 6. Here the acceleration a is upward. In a, at time t, the ship of length L is moving downward with a velocity v. In b, a time L/c later, the ship is stationary and the observer at the tail-end of the ship is right next to our stationary observer O. Let us assume that the relativistic Doppler shift accounts for the frequencies observed, and explore the consequences.

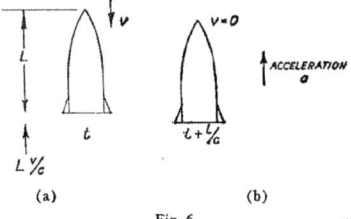

Fig. 6.

As we are dealing with moderate accelerations and velocities, we will assume Newtonian laws of motion. Further, we can at an appropriate point neglect Lv/c in comparison with L.

The radiation that the stationary observer, and the other stationary observer on the tail of the ship, see at time $t+L/c$ left the nose a time very nearly L/c earlier, when the ship was traveling downward with a velocity v. Hence, the observed relativistic Doppler frequency for this radiation will be

$$f_d = f \frac{\sqrt{1 + v/c}}{\sqrt{1 - v/c}} \doteq f(1 + v/c). \qquad (50)$$

Thus, the stationary observer and the observer in the tail of the ship find radiation as if from a clock running faster by a fraction $\Delta f/f$

$$\Delta f/f = v/c. \qquad (51)$$

Now, assuming Newtonian acceleration, very nearly

$$v = a \qquad t = aL/c. \qquad (52)$$

When we substitute v from (52) in (51) we obtain

$$\Delta f/f = aL/c^2. \qquad (53)$$

Einstein's principle of equivalence asserts that a field due to acceleration is indistinguishable from a gravitational field. A man cooped up in the space ship thus has no way of knowing that the ship is changing velocity. He may think of it as sitting on the ground and experiencing a gravitational field of acceleration a. Hence, we conclude that a clock at height L in a gravitational field of acceleration a runs faster than a clock on the ground by a fraction given by (53).

This is a well-known relativistic clock effect which predicts a red shift in radiation from a massive body such as the sun.

We can deduce this effect in another way. If a mass m is lifted to a height L against a field of acceleration a, it gains an energy maL and this is a gain of a fraction aL/c^2 of the total energy mc^2. Thus if a fraction of the mass is converted to photons which return to the ground a distance L below, these photons must have an energy greater by this same fraction aL/c^2 than if they had been emitted before the mass was raised. As frequency is proportional to energy, the frequency of the photons

from the source at a height L will be raised by just the fraction given by (53).

In connection with the "clock paradox" of Section IV, it has been objected by some that "relativity" requires that the same time elapse for the traveler as for stay-at-homes, but this is not so.

Special relativity can deal only with relative time rates ascribable to velocity differences, not with those caused by gravitational fields or accelerations. In Section IV we scrupulously took relative velocities into account. We insisted, however, that the ship change its velocity with respect to earth, star and the radiation as well. We did this because the rockets applied a force to the ship, but not to the earth, the stars and the radiation.

The ship captain might insist, however, that during acceleration his ship is being supported in a gravitational field, while all the heavens fall past it. In this case he would have to take into account the effect of this gravitational field in increasing the frequency of the radiation traveling toward him between his ship and earth. No doubt the complete and general relativist could do not only this, he could also treat a case of protracted acceleration and deceleration. It seems simplest, however, to proceed as in Section IV, and other results must be consistent with the results obtained there.

VI. Clock Rate on a Satellite

Let us imagine a very accurate oscillator on a satellite in a circular orbit at radius r. According to Newton's laws, the orbital velocity is given by

$$v^2 = gr(r_0/r)^2. \qquad (54)$$

Here r_0 is the radius of the earth and g is the acceleration of gravity at the surface. Accordingly, there is a fractional *reduction* in clock rate as observed from the earth of

$$1 - \sqrt{1 - (v/c)^2} \doteq (1/2)gr_0(r_0/r)/c^2. \qquad (55)$$

The potential energy of a unit mass at radius r with respect to the surface (at r_0) is

$$g(r_0/r)(r - r_0). \qquad (56)$$

In accordance with Section V, there is a fractional *increase* in clock rate because of the gravitation field of

$$g(r_0/r)(r - r_0)/c^2. \qquad (57)$$

Thus, the net fractional change of clock rate, $\Delta f/f$, is

$$\frac{\Delta f}{f} = \frac{g}{c^2} \frac{r_0}{r}\left(r - \frac{3}{2} r_0\right). \qquad (58)$$

We must note that this will be the frequency when the satellite is directly overhead; a Doppler shift will be present when the satellite is approaching or receding, in accordance with (37) (overhead corresponds to $\theta = 0$).

If the gravitational "red shift" were not taken into account, the frequency decrease at a radius of 6000 miles (a height of 2000 miles) would be about 4 parts in 10^{10}. When the red shift is taken into account the net frequency change at this altitude is, according to (58), zero.

We should note that in making calculations of this sort, in accord with Section V, we take into account velocities and gravitational fields in the observer's frame of reference. Fields experienced by someone else, as, for instance, a force felt in the satellite because of an acceleration of the satellite, are irrelevant.

VII. THE PHOTON ROCKET

Can we ever hope to travel with anything like the speed of light? For this purpose, some invoke the photon rocket. Of the total rest mass m_0 of the rocket we regard all but a fraction a (or a rest mass am_0) as fuel. We will not burn this fuel, nor fission it, nor fuse it. We will turn it all completely into radiation and use the pressure of this radiation to thrust the ship forward.

Of course we don't know how to turn matter completely into radiation. If we could do this, and if the minutest fraction of the radiation released heated the ship rather than pushing it forward, it would of course fry the crew. We will, however, assume that all the energy goes into pushing the ship, and none into heating it.

Before the ship sets off on its journey, it has zero velocity with respect to earth, a mass m_0 and a total energy m_0c^2. The ship shoots out radiation of total energy E_r to the left and itself shoots off to the right, attaining a velocity v. Its remaining rest mass is now am_0 and its relativistic mass is

$$\frac{am_0}{\sqrt{1 - (v/c)^2}}. \tag{59}$$

According to the conservation of energy, the energy of the radiation produced, E_r, plus the final energy of the ship must be equal to the initial energy of the ship, m_0c^2

$$E_r + \frac{am_0c^2}{\sqrt{1 - (v/c)^2}} = m_0c^2. \tag{60}$$

According to the conservation of momentum, the momentum of the radiation (directed to the left) must equal the momentum attained by the ship (directed to the right), or

$$E_r/c = \frac{am_0v}{\sqrt{1 - (v/c)^2}}. \tag{61}$$

From (60) and (61) we obtain

$$v/c = \frac{1 - a^2}{1 + a^2}. \tag{62}$$

Now, we remember that to make time pass only a tenth as fast on the ship as on the earth, we must attain 99.5 per cent of the speed of light; that is, v/c must be 0.995. If we use a photon rocket to attain this speed, what fraction of the initial mass or matter will we have left?

The answer turns out to be a fraction

$$a = 0.05. \tag{63}$$

This sounds extreme, but think of the trip out and back! We start out with a mass m_0. After getting up to 99.5 per cent of the speed of light we have left of the ship only a rest mass of 0.05 m_0. After stopping at the far end, we have only a mass (0.05) (0.05) m_0, or 0.0025 m_0. After starting back and then stopping at earth, we are left with only a fraction 0.00000625 of our original mass, which sounds rather impractical.

Did we do the wrong thing? Instead of using up rest mass to shoot radiation off to the left, we might have used it to shoot a part of the ship off to the left with a velocity v_1 and a relativistic mass m_1, while the ship proper, which retains a fraction a of the total original rest mass m_0, goes off to the right with a velocity v. Remember, m_1 is *relativistic* mass and am_0 is *rest mass*. With this in mind, the conservation of energy becomes

$$m_1c^2 + \frac{am_0c^2}{\sqrt{1 - (v/c)^2}} = m_0c^2. \tag{64}$$

The conservation of momentum is

$$m_1v_1 = \frac{am_0v}{\sqrt{1 - (v/c)^2}}. \tag{65}$$

From these equations we obtain

$$v/c = \sqrt{1 - a^2(1 + v/v_1)^2}. \tag{66}$$

We are of course interested in the case in which the fraction a is very small and v/c is very near to unity. In this case, unless the mass shot out to the left is very much greater than the remaining mass of the ship which moves to the right, v_1/c will be very nearly unity too. In other words, v/v_1 will be nearly equal to one. If this is true and is a is small, then (66) is approximately

$$v/c = 1 - 2a^2. \tag{67}$$

Under the same circumstances, according to (62) the photon rocket gives us a speed of approximately

$$v/c = 1 - 2a^2. \tag{68}$$

We see that (67) and (68) are the same. The remaining fraction a of the initial rest mass attains the same speed whether we turn all of the rest of the mass into photons and use these to accelerate the rest mass am_0, or whether we use some of the mass to shoot a part of the mass of the ship rearwards with a speed close to the speed of light.

We could do a little better if we could use the energy released to accelerate the ship by pushing against some fixed or very massive body—the universe, for instance. In this case all the energy will appear in the ship, and the conservation of energy will lead directly to the velocity, giving

$$\frac{am_0c^2}{\sqrt{1 - (v/c)^2}} = m_0c^2$$

$$v/c = \sqrt{1 - a^2} \qquad (69)$$

when a is very small, very nearly

$$v/c = 1 - a^2/2. \qquad (70)$$

We see that this is somewhat better than the other two cases [see (67) and (68)]. It is still not good enough to allow us to make a trip with nearly the speed of light. Moreover, I can't think of a plausible way to do it, except perhaps in taking off from a planet.

How is it that we find it possible on earth to accelerate electrons and protons far closer to the speed of light? The answer is, that we do have a massive body to push against, and besides this we do not rely on the energy of the particles, but instead have a huge fixed power plant; it is with these that we accelerate measly little atomic particles to speeds near that of light. With a sort of fixed space cannon and a huge mass-conversion power source we might shoot a space ship from earth with 99.5 per cent of the speed of light, but it seems unlikely that the crew would survive the acceleration.

VIII. Fuel from Space

Space is full, loosely speaking, of matter, which is mostly hydrogen. It is said that on the average there is about one atom of hydrogen per cubic centimeter, but to be on the safe side let us assume some larger number M of hydrogen atoms per cubic centimeter, or $M \times 10^6$ atoms per cubic meter. As the mass of a hydrogen atom is

$$1.66 \times 10^{-27} \text{ kilograms} \qquad (71)$$

the mass of hydrogen per cubic meter is

$$1.66 \times 10^{-21} M \text{ kilograms}. \qquad (72)$$

Let us assume a space ship of mass m_0 kilograms which, in literally tunneling through space, sweeps up all the matter over an area A, turns it into radiation, and uses the radiation to push the ship forward. In a distance L the ship has acquired and used a mass which we shall call m_i of interstellar matter; m_i will be given by

$$m_i = 1.66 \times 10^{-21} \, maL \text{ kilograms}. \qquad (73)$$

If v is the final velocity attained by the ship of rest mass m_0 and E_r is the energy of the radiation produced,

the conservation of energy tells us that

$$E_r + \frac{m_0c^2}{\sqrt{1 - (v/c)^2}} = (m_i + m_0)c^2. \qquad (74)$$

Initially, there is no radiation and neither the interstellar matter nor the ship has any momentum. Finally, in the tunnel behind the ship there is no interstellar matter (we used it up), there is radiation with momentum to the left, and the ship has a momentum to the right. The conservation of momentum tells us

$$E_r/c = \frac{m_0v}{\sqrt{1 - (v/c)^2}} \qquad (75)$$

From (74) and (75), we obtain

$$v/c = \frac{(1 + m_i/m_0)^2 - 1}{(1 + m_i/m_0)^2 + 1}. \qquad (76)$$

From (73) we easily see that if the distance of travel, L, is zero, m_i is zero, and thus (76) tells us that we are at a standstill.

How fast are we going after a ten-light-year trip? A light year is

$$\text{one light year} = 9.48 \times 10^{15} \text{ meters.} \qquad (77)$$

Let us assume a generous scoop (made of force fields, no doubt) which is 100 meters square, so that

$$A = 10^4 \text{ square meters.} \qquad (78)$$

We thus see that the mass of interstellar matter collected will be

$$m_i = (1.66 \times 10^{-21})(10^4)(10)(9.48 \times 10^{15})M$$

$$m_i = 1.57 \, M \text{ kilograms.} \qquad (79)$$

Surely, a habitable starship will weigh several thousand kilograms. Let us assume the mass of the ship to be 15,700 kilograms (about 17.5 tons). Let us generously assume not one but 1000 hydrogen atoms per cubic centimeter. This makes

$$m_i/m_0 = \frac{(1.57)(1000)}{(15,700)} = 0.1. \qquad (80)$$

From (80) and (76), we obtain

$$v/c = 0.093. \qquad (81)$$

Clearly, it is impossible to attain a velocity close to that of light by using interstellar matter as fuel.

30 August 1957, Volume 126, Number 3270

SCIENCE

Reprinted from Science
(Aug. 1957 - Pages 381-384)
By Permission

The "Clock Paradox" and
Space Travel

Edwin M. McMillan

In the pages of *Nature* there have appeared recently exchanges of correspondence between McCrea (1), Dingle (2), and Crawford (3). Dingle maintains that there will be no difference in age between a returned space traveler and his twin brother who stays home, while the other authors maintain that such a difference will exist, in just the amount computed by a straightforward application of the Lorentz transformation of special relativity. Dingle's argument is based on an old difficulty known as the "clock paradox," which stems from an apparent ambiguity in the answer to the question: If all motion is relative, how does one decide who traveled and who stayed home?

McCrea and Crawford have in my opinion clearly won the argument, and perhaps further remarks are superfluous. However, because of the considerable interest aroused, it may be worth while to restate the situation in new words and all in one place; hence this article. It has three sections. In the first, the "paradox" is stated and resolved, using only inertial coordinate systems; in the second, a treatment of accelerated coordinate systems based on the principles of special relativity is given; in the third, the possibility of practical implications for space travel is examined.

The "Paradox" and Its Resolution

The apparent difficulty in the "paradox" can be stated very simply. The twin brothers B and B' are in relative motion

The author is a professor of physics and member of the staff of the Radiation Laboratory at the University of California, Berkeley.

with the velocity v. If B considers himself at rest, he concludes from the usual formula for the time dilatation that the watch carried by B' will register a time interval

$$\sqrt{1 - (v^2/c^2)}\,\Delta t$$

while his own watch registers a time interval Δt. If B' makes a journey at velocity v and returns home at the same velocity, his elapsed time (as measured by his watch or by his own physiological aging) will be smaller by the factor

$$\sqrt{1 - (v^2/c^2)}$$

than the elapsed time experienced by B, who stayed home. However, if "all motion is relative," B can just as well say that he went on the trip while B' stayed at rest, and he can state that he is the one that should be younger. If both statements were correctly derived from the principles of special relativity, there would indeed be a paradox and one would conclude that the theory is not self-consistent and must be rejected. What we shall show is that the first statement is correct while the second is wrong; there is no true paradox, and the result that travelers live longer than stay-at-homes, while sometimes called "paradoxical," is really in the "strange but true" category.

To examine this matter more closely, we must set up the situation in greater detail. Suppose that B is at rest at the origin of an inertial coordinate system S, while B' has the same relation to an inertial coordinate system S', and that the two coordinate systems are in relative motion with a velocity v. Suppose also that the watches of the two brothers are set so that they each read zero when the two origins coincide, which defines the

starting point of the trip. The coordinates and times in the two systems are then related by the Lorentz transformation

$$x' = \gamma(x - vt) \tag{1}$$

$$t' = \gamma\left(t - \frac{v}{c^2}\,x\right) \tag{2}$$

$$\gamma = 1/\sqrt{1 - (v^2/c^2)}$$

From these equations, and from the fact that the brothers are located at their respective origins, we find for the relation between times observed at the location of B (where $x = 0$)

$$t' = \gamma t \tag{3}$$

while at the location of B' (where $x' = 0$), we find

$$t = \gamma t' \tag{4}$$

The quantities in these equations are clock readings t at coordinates x in S, and t' and x' similarly defined in S', while the equations give the relations between these quantities at localized events which are perceived in both systems. We may picture each system as carrying a long line of measuring rods and clocks, with observers to note down the clock reading and distance coordinate whenever an event occurs. These of course do not actually exist, but we are allowed to use them in discussion since the Lorentz transformation is constructed in such a way that all relations between events (such as the sending and reception of light signals) perceived where there are actual observers are consistent with observations that might be made elsewhere.

Now imagine a very simple event in which B' merely looks at his watch, notes that it has the reading τ, and also observes the reading of the particular clock in S that coincided with his location at that instant, obtaining the value $\gamma\tau$, as given by Eq. 4. This event alone is not sufficient to determine the relative elapsed time as seen by B and B'; it must be correlated with another event, which is a watch reading made "simultaneously" by B.

Here we run into trouble in the interpretation of the word *simultaneous*. First consider the situation from the standpoint of B. One of his corps of observers reports (at any time after the event) that he saw B' read his watch when he went by, and reports his readings of t and t', which are the same as those noted by B'

at the same time and place. B can then say: "When B' looked at his watch, he saw the reading τ; my own observer read the time as $\gamma\tau$ on his local clock; that clock is synchronized with my own watch; therefore, if I had looked at my watch when it indicated the time $\gamma\tau$, that event would have been simultaneous with the event consisting of B' looking at his watch." Second, consider the situation from the standpoint of B'. He says to his observers: "One of you was next to B when your clock read τ, which is the time when I read my watch. What did you see on the watch of B?" The result, from Eq. 3, would be τ/γ. Therefore, if B had looked at his watch when it read τ/γ, B' would have considered that event to be simultaneous with his own watch-looking event.

Thus, to one event at B', there correspond two different "simultaneous" events at B, depending on whose idea of simultaneity is accepted. This result is well known, and is not considered troublesome so long as B and B' are in relative motion at a distance apart; because of the symmetry of the situation, there is no way to decide who is "right," and there is no meaning in such a decision.

The usual next step is to have one of the observers reverse his motion, so that ages can be compared directly when the two come together again. However, the argument is just as cogent if we have them come to a state of relative rest, where age comparisons are meaningful even at a distance, and this approach is used because of its simplicity. The end of the trip will be defined by having one of the observers (say B') come to rest relative to the other.

Now an element of dissymmetry is introduced, since B' is the one who has to push the firing button of his decelerating rocket and whose accelerometer will deflect, while nothing will happen to B. This fact is used in a way which does not involve any arguments from general relativity (except insofar as it may give meaning to the concept of an inertial system) or any discussion of accelerated coordinate systems. The important thing is that, once B' has stopped, he has dissociated himself from S' and attached himself to S (with a shift in the zero point of his time scale, of course); the system S' remains unchanged. (We may note here that a reversal, rather than a stopping, of B' can be treated by having him transfer to a third inertial system S'' moving in the opposite direction; this more complicated procedure leads to a result for the round trip consistent with what we get for the outward part.) It is also postulated, as is usually done and as seems physically reasonable, that acceleration as such does not change his clock reading. His readings of τ on his own

watch and $\gamma\tau$ on the clock in S at which he stopped are, one may say, transferred bodily into S, and he and B can then compare notes with no ambiguity since they are at relative rest and since both use the same definition of simultaneity. Both B and B' agree that the ratio of elapsed times during the trip is that given by Eq. 4, and there is no "paradox."

Looking at the situation from another standpoint, one might inquire about the interpretation of a signal sent out by B' at such a time that its arrival at B is simultaneous (in S') with the stopping of B'. Such a signal, seen by B and interpreted as indicating "the end of the trip," would lead to a wrong result (to the usual "paradox" in fact). However, if B is more careful, he will realize the discrepancy in concepts of simultaneity and correct for it. From Eq. 2 we find that the difference in t between two events occurring at the same value of t' and at locations a distance Δx apart (in S) is equal to $v\Delta x/c^2$. The distance Δx in the case being considered is $v\gamma\tau$, the length of the journey in S; the events are the arrival of the signal at B and the stopping of B'. The time correction $v\Delta x/c^2$ has the value $(v^2/c^2)\,\gamma\tau$, or $[\gamma - (1/\gamma)]\tau$. When this is added to τ/γ, the observed value of t at the arrival of the signal, the value $t = \gamma\tau$ for the time of stopping of B' is obtained, in agreement with what was seen directly by observers in S. What will B' think of this? He sent the signal with the expectation that it would reach B when he reached the end of his journey, but when he gets into communication with the observers in S after the trip is over, they tell him that the signal arrived before he stopped. He resolves this apparent dilemma by realizing that when he jumped from S' into S he had to replace the concept of simultaneity in S' by that of simultaneity in S, the difference introduced thereby being equal to the "time correction" already defined.

At the cost of some repetition, we return to the question: How does one distinguish the stopping of B' from the starting of B, when the change in relative velocity is the same in both cases? Physically, the acceleration is felt by B' and not by B; mathematically, the acceleration is treated as a transfer from one inertial system to another, and here again the distinction is just as apparent. The essential symmetry of the treatment is clear if one notes that the concept of simultaneity is used in that inertial system in which *both* of the brothers are initially and finally at rest.

By a simple but interesting transformation, it is possible to relate the age difference of the brothers to their *changes* in velocity. For symmetry, allow both brothers to move, with velocities corre-

sponding to γ_1 and γ_2. Then the difference between their elapsed times t_1' and t_2' is given by

$$t_1' - t_2' = \int[(1/\gamma_1) - (1/\gamma_2)]dt =$$
$$[(1/\gamma_1) - (1/\gamma_2)]t - \int td[(1/\gamma_1) - (1/\gamma_2)]$$

Since we have postulated that $\gamma_1 = \gamma_2 = 1$ (that is, both brothers are at rest in the system in which t is measured) at the start and finish, the first term on the right vanishes. If the velocity changes are thought of as discontinuous (being expressed as changes in $1/\gamma$), the second term becomes a sum, giving:

$$t_1' - t_2' = \Sigma t\Delta(1/\gamma_2) - \Sigma t\Delta(1/\gamma_1)$$

This illustrates both the symmetry of the treatment and the importance of changes in velocity. Note that only the magnitudes of the velocities enter, and that the motion need not be confined to the x-axis.

Accelerated Coordinate Systems

It is of some interest, although not necessary for this problem, to inquire what happens in an accelerated coordinate system. In the treatment given in the preceding section, the system S' was not accelerated, but the traveler simply left it, like a man jumping off a train. Now we suppose that the train slows down with the man on it. The coordinate system (that is, the train, or a line of space ships if you prefer) carries measuring rods and clocks by which times and distances are measured. It is accelerated by applying power to the wheels or firing rockets at times considered simultaneous in itself. We start again with Eqs. 1 and 2 and suppose that the acceleration occurs suddenly at $t' = 0$. We now stand in S and watch the process. We note that the acceleration does not now seem to be simultaneous; Eq. 2 shows that, at each t, it occurs at an x given by $x = (c^2/v)t$. Thus the acceleration looks like a wave traveling with the velocity c^2/v. (The fact that this velocity is greater than c is not objectionable since the wave does not convey a signal.)

An acceleration wave in the form of a step of finite amplitude will not retain its shape since the wave velocity varies with v; therefore we deal at first with a step wave of infinitesimal amplitude dv. The new Lorentz transformation after the wave has passed is

$$x' = (\gamma + d\gamma)[x - (v + dv)t] \quad (5)$$
$$t' = (\gamma + d\gamma)\left[t - \frac{v + dv}{c^2}x\right] \quad (6)$$

where t' retains its meaning as a time considered simultaneous in the new S'. However, it no longer corresponds to the clock readings in the new S' except at

$x' = 0$. To see this, we note that a clock at a given x' does not change its reading suddenly when the acceleration occurs, while t' does make a sudden change. The latter is computed by subtracting Eq. 2 from Eq. 6, evaluated at the appropriate place and time in S—that is, at $t = (c^2/v)t$, with the result that t' at a given point changes suddenly by the amount

$$dt' = -\gamma x dv/c^2 = -\gamma x' dv/c^2$$

The clock reading at x', which we shall call T', does not suffer a corresponding instantaneous change, so that a difference between T' and t' appears:

$$d(T' - t') = \gamma^2 \frac{x' dv}{c^2} \quad (7)$$

A similar treatment applied to Eqs. 1 and 5 shows that the change in x' at a given point (say a given scale division on an actual measuring rod in S') is equal to zero. This is not trivial; x' is not defined as a scale reading but as a coordinate in terms of which the Lorentz transformation is valid; scale readings, like clock readings, cannot change suddenly, but their interpretation as coordinates can. For example, suppose that an acceleration were made to take place as a wave traveling with a velocity (in S) other than c^2/v. Then there would be a change in x' at a fixed scale reading in S'. The observer in S' would interpret this as an actual physical expansion or compression of his system arising from the fact that the acceleration no longer occurs simultaneously at all points in his system. In other words, the specification that the acceleration be simultaneous according to t' is necessary if we require that no "strains" be introduced into S'.

The concept of an acceleration wave with velocity c^2/v in S gives a simple pictorial representation of the generation of a Lorentz contraction with no discontinuous coordinate changes. Consider a measuring rod in S' with an apparent length l in S. The time interval in S for the wave to pass the length of the rod is $l/[(c^2/v) - v]$. During this interval, one end is moving faster than the other by the amount dv, and the apparent final change in length is

$$- l dv/[(c^2/v) - v] = -\gamma^2 l v dv/c^2$$

According to the Lorentz contraction $l = l'/\gamma$. Differentiating this, we find that

$$dl = -l' \gamma v dv/c^2 = -\gamma^3 l v dv/c^2$$

in agreement with the result obtained from the wave picture.

It is now of interest to find out whether Eq. 7 is consistent with general relativity, which says that two clocks, at a distance x' apart in a system suffering acceleration a', will appear to differ in rate by the amount $a'x'/c^2$, and therefore will acquire a difference in reading $\gamma' du/c^2$

if the acceleration continues just long enough to produce a velocity change du'. The velocity change du' is a small increment of velocity from rest, as seen by the observer in the accelerated system; to find the corresponding change in the velocity v as seen by an outside observer, we differentiate the relativistic formula for the composition of velocities, with the result that $dv = (1/\gamma^2) du'$. The change in clock readings, expressed in terms of dv, is therefore in agreement with Eq. 7, and we have derived the change in clock rates in an accelerated system using special relativity alone. The generalization to a gravitational field requires, of course, the use of the equivalence principle of general relativity.

Now we wish to consider the effect of a finite change in v, which can be thought of as taking place by means of a set of superimposed infinitesimal stepwise acceleration waves. The resulting composite wave can be stepwise at only one point because the wave velocity varies with v. In the case of a stepwise increase in v (assumed to be initially positive) at $x = 0$ and $t = 0$, the wave will spread out in the forward direction as t increases, and cannot be defined for $x < 0$, $t < 0$. In the case of a decrease in v at the same place and time, the wave is spread out behind at earlier times, and cannot be defined for $x > 0$, $t > 0$. In order to generate such a wave, the firing of the rockets in the line of space ships defining S' must be scheduled differently at different points; if the acceleration is instantaneous at one point, it must be spread out, according to local clock readings, and the rate of acceleration must be correspondingly reduced, at other points. This corresponds to the fact that each velocity increment must occur simultaneously in t', while the clocks do not continue to indicate t'.

After the complete acceleration wave has passed, the total change in clock readings is obtained by integrating Eq. 7. If the relative velocity changes, for instance, from 0 to v, we find

$$T' - t' = \int_0^v \gamma^2 \frac{x'}{c^2} dv = \frac{x'}{c} \tanh^{-1}\left(\frac{v}{c}\right) \quad (8)$$

This is not to be confused with the "time correction" used earlier, which has the same value only when $v \ll c$. The "time correction" is the result of a discrepancy in the estimation of simultaneity in *two different coordinate systems* with relative motion; Eq. 8 represents a desynchronization of clocks in *the same coordinate system*.

Finally, we inquire what relation this has to the "paradox." When B' stops, suppose that he brings his whole coordi-

nate system to rest with him, and then compares the clock in his system at the location of B with the watch of B. This clock is out of synchronization with his own by the amount shown by Eq. 8, but he is presumed to be clever enough to know this (or to find it out by measurement after he has stopped), so he corrects for it, and the actual value of Eq. 8 does not enter into the result. He might just as well have used the already existing system S, in which he is at rest after stopping, and simply taken the reading of the clock opposite which he stopped to be elapsed time as estimated by B, which leads to the same result and is equivalent to the choice of Eq. 4 for the relation between the elapsed times.

Implications for Space Travel

The great recent interest in the "clock paradox" is based on its possible consequences for travelers in space; this transcends in popular appeal the more basic matters of principle involved. Therefore let us look at it from that standpoint. First, could an acceleration tolerable to human beings, acting for a reasonable time, produce velocities great enough to give an appreciable time dilatation? To answer this, we must find the distance x (in the rest system of the starting point) traveled in time t' (in the traveler's rest system) while the traveler feels the constant acceleration a'. We first find the velocity v by integrating the relation $dv = (1/\gamma^2) du'$ (introduced in the preceding section), noting that $du'/dt' = a'$, with the result

$$v/c = \tanh \frac{a'}{c} t'$$

Then

$$x = \int_0^t v dt = \int_0^{t'} \gamma v dt' = \frac{c^2}{a'}\left(\cosh \frac{a'}{c} t' - 1\right)$$

If x is measured in light years, t' in years, and a' in units equal to the normal gravitational acceleration at the surface of the earth, then

$$x = \frac{0.97}{a'}\left(\cosh \frac{a'}{0.97} t' - 1\right)$$

This is a remarkable result. For example, a man traveling for 21 years under a constant acceleration of 1 g would go a distance of 1.2×10^9 light years! If he then reversed his acceleration, he would finally come to rest at a point 2.4×10^9 light years distant from his starting place, having spent 42 years of his life on the trip. The results are less dramatic for shorter trips; even for rather high accelerations (on a human scale), the relativistic time modifications are negligible for travel within the solar system. For example, a man going to Neptune and stopping there, at an acceleration of 10 g, would spend 5 days on the trip but would gain only 1.5 minutes of time.

Then there is the question of the energy involved. The man who travels for 21 years at 1 g reaches a value of γ equal to 1.2×10^9, at which point his kinetic energy is utterly fantastic. If his vehicle weighs (at rest) 1 ton, then its energy content is equal, in round numbers, to the energy released in the annihilation of 10^9 tons of matter, or in the fission of 10^{12} tons of uranium; it would be sufficient to melt the entire crust of the earth to a depth of about 30 miles. The man who makes the more modest trip to Neptune at 10 g reaches $\gamma = 1.0025$, and the kinetic energy of his 1-ton ship (2×10^{17} joules) corresponds to that released in the fission of about 2 tons of uranium; because of the limited efficiency of rocket propulsion, the actual energy needed would be much greater. The use of such energy quantities in a rocket ship is so far beyond any foreseeable practical limits, and the time gain in that case is so small, that it is hard to picture a practical case of space travel in which the time dilatation can be considered important. This conclusion, of course, does not detract from the interest of the fundamental principles involved in the "clock paradox" (4).

References and Notes

1. W. H. McCrea, *Nature* 167, 680 (1951); 177, 784 (1956); 178, 681 (1956); 179, 909 (1957).
2. H. Dingle, *Nature*, 177, 782, 785 (1956); 178, 680 (1956); 179, 865, 1242 (1957).
3. F. S. Crawford, Jr., *Nature* 179, 35, 1071 (1957).
4. I would like to express my appreciation to Frank S. Crawford, Jr., David L. Judd, W. K. H. Panofsky, Henry P. Stapp, and Edward Teller for many useful discussions. Since I have not read widely in the literature of this subject, I apologize to any authors who may have already published any of the material given.

Relativistic Observations and the Clock Problem (*).

J. TERRELL

Los Alamos Scientific Laboratory, University of California - Los Alamos, N. Mex.

(ricevuto il 21 Dicembre 1959)

Summary. — Relativistic observational data are discussed, with the purpose of clarifying some aspects of the clock problem, usually called the « clock paradox » or « twin paradox ». Einstein's position, that an ideal clock which moves in a closed curve with respect to an unaccelerated clock will indicate the passage of less time, is supported. It is pointed out that the sets of observational data of two observers who take the place of the clocks in the above situation will not be at all similar. Furthermore, the data of the accelerated observer, obtained by means of single Doppler shift and visual observational methods, will be highly implausible and inconsistent with data obtained by radar and double Doppler shift methods. Thus the accelerated observer will be under no temptation to consider himself in a situation equivalent with that of the unaccelerated observer, and should not be surprised to discover upon returning that he has aged less than the other observer. It is pointed out that the accelerated observer will see striking effects due to relativistic aberration which will not be seen by the other observer, but that neither observer will be able to see or photograph the Lorentz contraction. Only the special theory of relativity is necessary in these calculations, since no genuine gravitational fields, produced by massive bodies, are involved.

1. – Introduction.

More than fifty years ago, in his original paper on the special theory of relativity (¹), EINSTEIN stated that an ideal clock which is moved in a closed curve with respect to an unaccelerated reference system will be found, at the end of its journey, to have lost time with respect to a clock stationary in that.

(*) Work performed under the auspices of the U.S. Atomic Energy Commission.
(¹) A. EINSTEIN: *Ann. Phys.*, **17**, 891 (1905).

Reprinted from Nuovo Cimento
(May 1960 - Pages 457-468)
By Permission

system. This statement is usually called the « clock paradox » or « twin paradox » (in reference to accelerated and unaccelerated observers). During the last-twenty years the clock problem has been much discussed by DINGLE ([2]), who believes that EINSTEIN made a « regrettable error », and by many other writers who have responded in defense ([3-23]) of Einstein's statement. All standard textbooks on relativity and innumerable papers not mentioned here have supported Einstein's view; however, a few writers ([24-26]) appear to have accepted Dingle's views.

The basic reasons for the continuing controversy include the following:

1) The use of the word « paradox », which implies a contradiction between two correct statements and leaves the impression that the matter is not well understood. It is probably unfortunate that the clock problem has been referred to as an « apparent paradox ».

([2]) H. DINGLE: *Nature*, **144**, 888 (1939); **145**, 427, 626 (1940); **146**, 391 (1940); *The Special Theory of Relativity* (New York, 1941; also later editions); *Amer. Journ. Phys.*, **10**, 203 (1942); **11**, 228 (1943); *Nature*, **177**, 782 (1956); **178**, 680 (1956); *Proc. Phys. Soc. (London)*, A **69**, 925 (1956); *Nature*, **179**, 865, 1129, 1242 (1957); **180**, 500, 1275 (1957); *Austral. Journ. Phys.*, **10**, 418 (1957); *Science*, **127**, 158 (1958); *Bull. Inst. Phys.*, **9**, 314 (1958).

([3]) J. W. CAMPBELL: *Nature*, **145**, 426 (1940).

([4]) F. C. POWELL: *Nature*: **145**, 626 (1940).

([5]) P. S. EPSTEIN: *Amer. Journ. Phys.*, **10**, 1, 205 (1942).

([6]) L. INFELD: *Amer. Journ. Phys.*, **11**, 219 (1943).

([7]) W. H. McCREA: *Nature* **167**, 680 (1951); **177**, 784 (1956); **178**, 681 (1956); *Proc. Phys. Soc. (London)*, A **69**, 935 (1956); *Nature*, **179**, 909 (1957); *Discovery*, **18**, 175 (1957).

([8]) H. E. IVES: *Nature*, **168**, 246 (1951).

([9]) G. THOMSON: *The Foreseeable Future* (Cambridge, 1955), pp. 88-89.

([10]) F. S. CRAWFORD, Jr.: *Nature*, **179**, 35, 1071 (1957).

([11]) S. F. SINGER: *Nature*, **179**, 977 (1957).

([12]) W. COCHRAN: *Nature* **179**, 977 (1957); *Proc. Camb. Phil. Soc.*, **53**, 646 (1957).

([13]) J. H. FREMLIN: *Nature*, **180**, 499 (1957).

([14]) C. G. DARWIN: *Nature*, **180**, 976 (1957).

([15]) J. D. ROBINSON and E. FEENBERG: *Amer. Journ. Phys.*, **25**, 490 (1957).

([16]) E. M. McMILLAN: *Science*, **126**, 381 (1957); **127**, 160 (1958).

([17]) R. M. FRYE and V. M. BRIGHAM: *Amer. Journ. Phys.*, **25**, 553 (1957).

([18]) G. BUILDER: *Austral. Journ. Phys.*, **10**, 246, 424 (1957); **11**, 279 (1958); **12**, 300 (1959); *Phil. of Science*, **26**, 135 (1959); *Bull. Inst. Phys.*, **8**, 210 (1957).

([19]) C. B. LEFFERT and T. M. DONAHUE: *Amer. Journ. Phys.*, **26**, 515 (1958).

([20]) A. D. FOKKER: *Physica*, **24**, 1119 (1958).

([21]) R. H. ROMER: *Amer. Journ. Phys.*, **27**, 131 (1959).

([22]) E. FEENBERG: *Amer. Journ. Phys.*, **27**, 190 (1959).

([23]) C. C. MacDUFFEE: *Science*, **129**, 1359 (1959).

([24]) E. L. HILL: *Phys. Rev.*, **72**, 236 (1947).

([25]) L. ESSEN: *Nature*, **180**, 1061 (1957).

([26]) E. G. CULLWICK: *Electromagnetism and Relativity* (New York, 1957).

2) The incorrect belief that the statement «all motion is completely relative» is consistent with the special theory. Associated with this belief is a philosophical reluctance to accept relativistic aging differences. These are Dingle's basic tenets. However, the special theory predicts symmetry of observation for unaccelerated observers only; this symmetry is by no means true of other reference frames. The general theory of relativity makes no statements about symmetry of observations by two observers.

3) The incorrect belief that the clock problem cannot be handled by the special theory because of the accelerations. Accelerations were treated by use of the special theory in Einstein's first paper ([1]). Apparently it is now recognized by most writers, including DINGLE, that the general theory is not necessary in this matter. Both MØLLER ([2]) and MCMILLAN ([16]) have used special relativity to treat continuous acceleration in the clock problem. The general theory would be necessary only if genuine gravitational fields produced by massive bodies were involved.

4) Insufficient, and in some cases inaccurate, discussion of what the two observers («twins») will be able to observe. It appears that this matter has not been thoroughly considered. For instance, it was pointed out only very recently ([26]) that neither observer would be able to see or photograph the Lorentz-Fitzgerald contraction.

It is the purpose of this paper to discuss, in more detail than has previously been given, precisely what observations could be made by the accelerated and unaccelerated observers. It is hoped that this will make it clear in what ways the two observers are not equivalent.

2. - Unaccelerated case.

The simplest clock problem which has figured in the controversy involves three observers, A, B, and C, none of whom is accelerated; all are moving with uniform relative velocities along a single straight line in field-free space. In order to avoid discussing infinite sets of observers in each co-ordinate system, it will be assumed here and in other cases that each observer is equipped with radar, radio, and optical equipment for the purpose of measuring distances and relative velocities and synchronizing clocks. The radar method of measuring distance does not differ in principle from the optical methods discussed by EINSTEIN ([1]); it involves a signal moving at the velocity of light c, which is transmitted by observer A to observer B (for instance) and either reflected or retransmitted back to A. Observer A then plots the distance to B as $c(t_r - t_t)/2$ at the time $t_A = (t_r + t_t)/2$, where t_t and t_r are the times of trans-

---

([27]) C. MØLLER: *The Theory of Relativity* (Oxford, 1955), p. 258.

mission and reception as measured by A. In addition, B transmits to A the reading of B's clock at the time the signal is received by B (telescopic observation by A would give the same result), and A plots this reading (t_s) also for the time t_A.

It is postulated that the situation as observed by A with his equipment is this: B passes him with relative velocity r at the time $t_A = 0 = t_s$. synchronizing his clock in passing. Observer B attains a distance L at time $t_A = L/r$ and simultaneously passes observer C, travelling in the opposite direction, also with velocity r relative to A. Observer C synchronizes his clock to that of B at the moment of passing: there is no difficulty in doing this when the distance between them is vanishingly small. Finally C passes the position of A at $t_A = 2L/r$ and A and C compare clocks.

The result is unambiguous if the postulates of special relativity are accepted. Observer A observes the clocks of both B and C to be running uniformly slower than his, with the relation being $t_s = t_c = xt_A$ according to the data of A, where $x = \sqrt{1 - r^2/c^2}$. He will thus observe the reading $t_c = 2xL/r$ on C's clock at the moment of passing, when $t_A = 2L/r$. It cannot be correctly argued that A would observe C synchronizing his clock with that of B, but somehow making a mistake in the process (as observed by A) of just enough to make $t_c = t_A$ at the moment when A and C pass. If this were so, another observer with zero velocity relative to A, located at the point where B and C pass, would have to observe the same thing, and this is not possible if B and C see their synchronization as correct. It is a standard result in the special theory of relativity that two observers in the same inertial system, even though widely separated, can synchronize their clocks with no difficulty and will then obtain exactly the same observations of phenomena occurring in other systems.

Since A, B, and C are all assumed to be in inertial systems of co-ordinates, there is no reason to prefer the observations of A to those of B or C. Observer B, for instance, observes A's clock to be too slow by the same factor, x, that A observes for B's clock, and observes C's clock to be even slower than that of A, since B measures C's velocity to be $2r/(1+r^2/c^2)$, which is greater than r. Hence B agrees that C's clock reads less than that of A when C passes A. Similarly, the data of observer C indicate that A's clock is slow by the factor x, but that B's clock is even slower, thus accounting for the fact that C's clock, having been synchronized to that of B, still reads less time than A's when A passes C.

3. – Accelerated case.

In the situation just described, A, B, and C are equally good as observers. Their observations are different, but this fact is a basic phenomenon of special relativity. As long as they continue their uniform, unaccelerated, velocities, there is no basis for saying that anyone's clock is « really » indicating the passage of less time than another's clock; to do so would be to give preference to one of the three co-ordinate systems, However, such a statement does become possible if the observers are not in equivalent situations. The unaccelerated situation becomes the familiar clock problem if one change is made, substituting observer *B* for observer *C* at the time of their meeting. Thus we deal only with observers *A* and *B*; as observed by *A*, *B* synchronizes his clock in passing, travels with velocity *r* to a distance *L*, then reverses direction in a time negligible with respect to *L/r* and returns to the position of *A*. The observations of *A* are essentially the same as in the unaccelerated case, so that *B*'s clock will read less than *A*'s upon the second meeting. The observations of *B*, who does not remain in a single inertial system, will be confusing and apparently internally inconsistent, as will be discussed later. Thus the acceleration which *B* undergoes makes a real difference in the status of the two observers.

The acceleration period of observer *B* has been assumed to be short enough that it need not be considered in the calculations. The clock problem may be solved for lengthy periods of acceleration, but the calculations are also more lengthy ([16,27]), though straightforward. In no case does it become necessary to use the general theory, although the clock problem may be discussed from this point of view ([17,19 27 29]) if desired. In all calculations it is necessary to assume that the clocks are ideal clocks, which run during acceleration at the same rate as unaccelerated clocks with respect to which they are momentarily stationary. Actual clocks may not be ideal; they may, for instance, stop permanently under acceleration. Neither the general nor special theories of relativity can predict fully the behavior of actual clocks ([30]).

We thus have to assume, if the problem deals with actual clocks, that *B*'s clock does not change reading in an essentially discontinuous way during the short acceleration period. It is not physically reasonable that another observer, stationary with respect to *A* and located in the small area of *B*'s acce-

---

([2b]) J. TERRELL: unpublished paper on *The Clock « Paradox »* (1957); *Bull. Am. Phys. Soc.*, **4**, 294 (1959); *Phys. Rev.*, **116**, 1041 (1959).

([29]) R. C. TOLMAN: *Relativity, Thermodynamics, and Cosmology* (Oxford, 1934) p. 194.

([30]) The assumption of ideal clocks, or that actual clocks may be so corrected as to substitute for ideal clocks, is essential to the treatment of acceleration in either the special or general theories. However, accelerations observed in studies of fundamental particles can be so great that no conceivable clock would be ideal, as even the fundamental particles are disrupted. Without a complete theory of the constitution of such particles no completely accurate description can be given for such high accelerations. With lesser accelerations fundamental particles should be exceedingly good approximations to ideal clocks.

leration, would observe such a jump in B's clock reading. It would be completely unreasonable to assume that the jump would be just enough to make the clocks of A and B coincide at the end of the journey, since the jump would then depend on the distance of A from the additional observer. The third observer has been momentarily introduced merely to forestall doubts about observations at a distance; according to the special theory his observations would be identical with those of A, after the usual corrections for time lags.

CRAWFORD [10] has pointed out that the assumption of no jump in clock reading is experimentally correct for μ-mesons (one type of physical « clock »). Cosmic-ray data indicate [31-33] that μ-mesons have a much longer observed lifetime at relativistic velocities than at rest, as expected, and that the ages of μ-mesons which come to rest at different altitudes above sea level are not equalized by any anomalous disintegration rates during deceleration.

The difference in clock readings of observers A and B may be discussed very simply in terms of Minkowski's four-dimensional space-time. One of the basic results of special relativity is that the four-dimensional interval s between two events, defined by $s^2 = c^2 t^2 - \sigma^2$, is an invariant quantity, the same for all unaccelerated (« Galilean » or inertial) co-ordinate systems. In this equation t and σ are the measured intervals in time and space. If the departure, turnaround, and return of B are designated by subscripts 1, 2, and 3, the interval between events 1 and 2 (or between 2 and 3) is given, in terms of A's data, as $s_{12} = s_{23} = \sqrt{(cL/r)^2 - L^2} = (Lc/r)\sqrt{1 - r^2/c^2}$. Since observer B is physically present at events 1 and 2 (and is in an inertial system during this time), he will measure the interval as entirely timelike, the time interval being $\tau_{12} = (L/r)\sqrt{1 - r^2/c^2}$. He obtains the same result for τ_{23}; thus he measures the time interval for his entire journey as $(2L/r)\sqrt{1 - r^2/c^2}$, while observer A measures it as $2L/r$, a longer time.

It is true in general that an observer who is physically present at two events (and unaccelerated during this interval) will measure less time between the two events than any other observer who is in a different inertial system and hence observes a space interval between the two events. This statement, of course, follows immediately from the definition of the invariant space-time interval s.

4. – Data of the two observers.

The non-equivalence of observers A and B becomes quite clear when their sets of data on measured distances and times are compared, as McCREA [7]

[31] B. ROSSI, N. HILBERRY and J. B. HOAG: Phys. Rev., 57, 461 (1940).
[32] F. RASETTI: Phys. Rev., 60, 198 (1941).
[33] H. TICHO: Phys. Rev., 72, 255 (1947).

has done. In order to make the situation slightly more general, it is now assumed that B comes to rest with respect to A and remains at rest for a time $\varDelta$ before beginning the return trip. This situation then covers several different types of clock problems simultaneously; the simpler problem discussed in the previous section is, of course, obtained by setting $\varDelta = 0$.

We postulate, then, that A, making radar measurements of distance as a function of time, observes B to synchronize his clock in passing, proceed at velocity v to a distance L, pause at this distance for a time $\varDelta$, and return at velocity v. He also observes B's clock to be losing time during the periods of relative motion, so that at the second meeting B's clock reads $2xL/v - \varDelta$, while A's clock reads $2L/v + \varDelta$. What B observes, using similar equipment, may be calculated by use of the Lorentz transformation or by the elementary

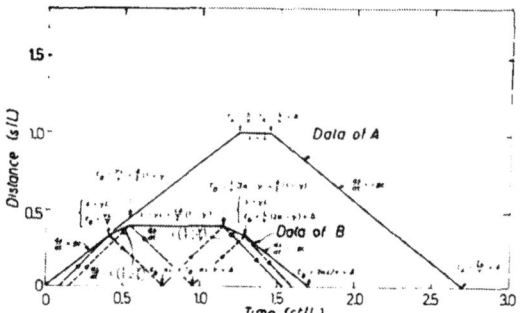

Fig. 1. – Data of observers A and B on relative distance as a function of time, obtained by radar or double Doppler shift methods, and in the case of observer A also by the use of single Doppler shift or visual observations. Paths of a few light or radar signals as plotted by B are shown as dashed lines. In this and the other figures it is assumed for illustrative purposes that $\beta = v/c = 0.8$, $x = \sqrt{1-\beta^2} = 0.6$, $\gamma = \sqrt{(1-\beta)/(1+\beta)} = \frac{1}{3}$, and $\varDelta = 0.2L/c$.

process of tracing the interchange of signals, and is shown in Fig. 1. For the purpose of drawing the figure the choices $\beta = v/c = 0.8$ and $\varDelta = 0.2L/c$ have been made, so that $x = \sqrt{1-\beta^2} = 0.6$ and the Doppler shift factor $\gamma = \sqrt{(1-\beta)/(1+\beta)} = \frac{1}{3}$. This numerical value of relative velocity is also the one chosen by DARWIN [14], to avoid « tiresome irrationalities ». The postulated observations of A are also shown, and it is seen that the two sets of observations are not at all similar.

The radar measurements of distance which B makes during the initial stage of his journey, with transmission and reception both occurring before B undergoes acceleration, indicate a relative velocity c, the same result obtained by A. The same result is also obtained during the last stage of B's journey. Those radar measurements for which reception (or transmission) occurs during the «pause» of time Δ indicate a smaller relative velocity given by $c(1 - \gamma)/(1 - \gamma)$. Those radar measurements made by B for which transmission occurs on the outward journey and reception occurs on the return trip take a constant time, as measured by B, and indicate that A, is for a time, at a constant distance less than L, given by $\gamma L - \Delta(1 - \gamma)/2$. It is assumed here that $\Delta \leqslant 2L/c$, so that B could not send and receive the same radar signal during the «pause» (for larger values of Δ observer B would observe the distance of A to increase continuously to the value L and then remain constant for a time).

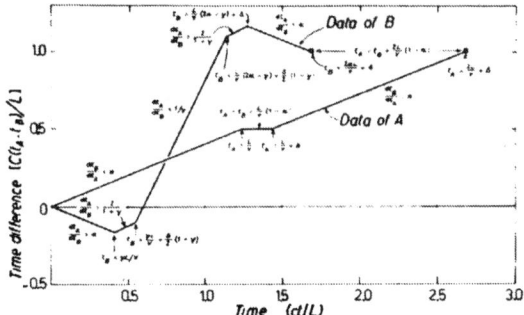

Fig. 2. – Data of observers A and B on differences in clock readings as functions of time. Both sets of data have been corrected by means of the distance observations of Fig. 1.

Fig. 2 shows the two sets of data on the difference in readings of the two clocks, as observed after correction for the time which light (or radio) signals take to reach the observers. As in the case of the distance measurements, those measurements completed within the first part (or the last part) of B's journey give the usual result, that B observes A's clock to be losing time the ratio of rates being α. However, the data received by B during the «pause» show A's clock to be gaining time at a rate given by $dt_A/dt_B = 2/(1 + \gamma)$; this is also true for data received at a later time, corresponding to radar transmission during the time Δ. Moreover, for the first part of B's return journey the incoming corrected time data of B indicate that A's clock is running extremely fast, by the factor $1/\gamma$. Naturally, these points are automatically plotted for an earlier time than that of reception of data. Finally, B observes A's erratic clock to be ahead of his on the occasion of reunion, which is at a time $2L_j/c + \varDelta$ for A and $2xL/c + \varDelta$ for B.

It is possible for observers A and B to take additional data on their relative velocities by measuring the Doppler shift of radar signals. If they observe the frequency f_r of their own radar signals (originally at frequency f_t) as returned to them by reflection, special relativity predicts that this «double Doppler shift» will be given by $f_r/f_t = \gamma^{\pm 2}$ for inertial reference frames: the minus sign is to be used for the case of decreasing distance between the observers. The velocity may be calculated from the relation $r/c = (1 - \gamma^2)/(1 + \gamma^2)$, and the distance may be obtained by integration of velocity. If observer A chooses to measure the Doppler shift of the radar signals transmitted by observer B, he must use the equation $f_r/f_0 = \gamma^{\pm 1}$ and solve for r in the same way. Here the negative exponent is once again to be used for the case of relative motion toward the observer: f_0 is the transmitted frequency as measured in the rest system of the transmitter, presumed to be known to A.

Observer A would obtain exactly the same set of data in either of these ways as with his radar measurements. Observer B would verify his radar data by measurement of the Doppler shift of his own reflected signals, but from single Doppler shift data would obtain a conflicting and nonsensical set of data as to velocity, which would, however, be consistent with his visual observations, to be discussed. He could obtain agreement between both types of Doppler data only by first correcting the rest-calibrated frequency f_0 of A's transmitter to a presumed frequency f_t as measured in B's co-ordinate system, using his data on the rate of A's clock, and then applying the appropriate equation ([34]), $f_r/f_t = (1 \pm r/c)^{-1}$. On this last basis his single Doppler shift data would give him no independently useful data, requiring both radar distance and clock observations for corrections.

It has become apparent that observer B is in trouble if he attempts to apply the simple equations given by special relativity for unaccelerated observers, and most of his difficulties have not yet been mentioned. If he should have the point of view that his situation is precisely equivalent to that of observer A, his confidence might be shaken by the strong accelerations which he, and not observer A, will feel. In any case, B observes at these times a sudden and permanent (until the next acceleration) change in the apparent

---

([34]) This equation is correct relativistically, and may be applied by observer A without difficulty. The relativistic Doppler shift may be derived as the product of the non-relativistic shift for stationary observer, as given here, and the relativistically reduced rate of a moving clock. A similar statement may be made for the assumption of stationary transmitter.

directions of stars and other external objects, a shift which A does not see. This is the optical phenomenon of aberration, the expression for which, derived [1] from the Lorentz transformation, is $\sin \theta = (x \sin \theta')/(1 - \beta \cos \theta')$. Here θ is the polar angle of an object as seen by any observer, $i.e.$, the apparent visual direction of the object, and θ' is the apparent direction of the same object as seen simultaneously by another observer at the same location but moving in the direction $\theta = 0 = \theta'$ with relative velocity $r = \beta c$. If an observer at the point of acceleration, stationary with respect to A, sees a star at $90°$ to the path of B, for example, B will see it somewhat ahead at about $37°$ (for $\beta = 0.8$; in general the angle would be given by $\sin \theta = x$) during the first part of his journey, at $90°$ while he is at rest with respect to A, and at $143°$ during the last part of his trip.

It should be noted, however, that neither observer A nor B will see or photograph any Lorentz-Fitzgerald contractions of stars or groups of stars at any time. As has been recently pointed out ([25]), the contraction due to relativistic motion is optically invisible. In effect, aberration is equivalent to a conformal transformation on the surface of a sphere centered at the point of observation, on which are plotted the apparent directions of distant objects. If a distant object is seen by one observer to have a particular outline, circular for instance, it will be seen with precisely the same outline (for sufficiently small subtended angle ([25])) by any other observer in relative motion but simultaneously at the same point. He will, however, see it at a different angle and apparent distance (the ratio of apparent distances is just the Doppler shift ratio), and there will be curious distortions of perspective, or of appearance with the use of stereoscopic vision ([26]). The visual data of any observer will lead to the well-known contraction only when corrected for the finite velocity of light.

If B observes A through a telescope, he will see sudden changes in the apparent distance of A at the times of acceleration; that is, the angle subtended by the face of A's clock or meter stick will suddenly change. The aberration equation yields the result that, when B comes to rest with respect to A, B observes A suddenly to shrink in apparent subtended angle, the ratio of change being γ. The effect is that A, apparently at distance γL, suddenly appears to be at distance L. As soon as B begins to move toward A (from the point of view of A), B observes a further apparent shrinkage of A, the ratio again being γ, so that A now appears to be at a distance L/γ. Observer B, knowing the finite velocity of light, naturally ascribes these events to times

---

([25]) R. PENROSE: *Proc. Cam. Phil. Soc.*, **55**, 137 (1959), has recently shown that for the special case of a circular outline there is no restriction as to subtended angle. Spheres will be seen as having circular outlines by all observers, regardless of relative velocities and subtended angles.

earlier than those of observation. By observing the angle subtended by A throughout his journey, B has a perfectly valid way of plotting the distance of A as a function of (corrected) time. That is, the method is valid for an inertial system of co-ordinates, and A would merely verify his radar measurements in this way. However, B's data, shown in Fig. 3, will give B the

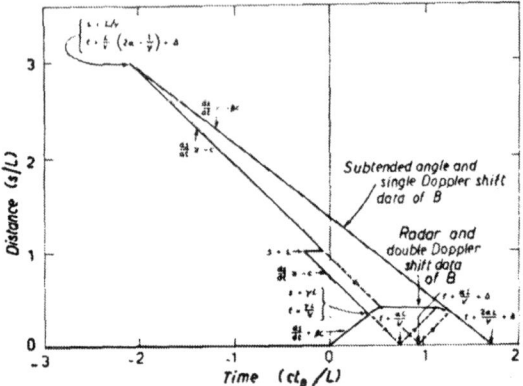

Fig. 3. – Data of observer B on the distance of A, as taken in several different ways. Paths of a few light or radar signals as plotted by B are shown as dashed lines.

problem of deciding how A can move at speeds slightly greater than the speed of light and travel alternately backward and forward in time, thereby existing in three or occasionally five places simultaneously. Observer B will also have to decide why these data differ from his radar and double Doppler shift data; as mentioned earlier, his single Doppler shift data could be interpreted as being consistent with his visual observations.

The simplest solution to the difficulties facing observer B would be that he decide that he has changed velocity and has not remained in a single inertial system. He can then correct his data to obtain a consistent set for some unaccelerated co-ordinate system. An alternative which B could choose is to apply the general theory of relativity and assume himself to have been unaccelerated throughout these events. He could, because of the principle of equivalence, account for all observed effects by introducing moving gravitational fields having the nature of plane shock waves, which passed him coincidentally with his application of rocket power (or whatever produced his accelerations as observed by A). Thus observer B could assume himself to have remained stationary while A and the rest of the universe were accelerated.

However, this would be a somewhat complicated, coincidental, and physically implausible explanation, although it would agree with the general theory of relativity and would account for all the observations discussed.

5. – Conclusion.

It is the intent of this paper to clarify some features of the clock problem, in particular what the accelerated observer could actually observe by the use of conceivable equipment, and to show that in all aspects the problem can be handled by the tools of special relativity alone. The result is that the accelerated observer will definitely observe less elapse of time than the unaccelerated observer. No paradox, in the sense of a contradiction, is involved since the two observers cannot be considered, and cannot consider themselves, to be in equivalent situations.

* * *

I am indebted to many persons at this Laboratory for informative discussions, and in particular to Drs. A. W. SCHARDT and A. M. LOCKETT for careful examination of some of the points discussed in this paper. I am grateful to Drs. C. G. DARWIN, W. H. McCREA, J. A. WHEELER, and H. A. WILSON for helpful advice, and to Dr. H. DINGLE for pointing out some ambiguities of language in an earlier version of this paper (with which he disagrees).

On Solutions of the Clock Paradox

G. David Scott

University of Toronto, Toronto, Ontario

(Received March 23, 1959)

The nature of the clock paradox is discussed and three solutions are referred to: (a) length contraction, (b) Doppler effects, (c) world lines in chronogeometry. The Special Theory of Relativity gives a complete explanation of the problem and it is pointed out how the use of the General Theory provides merely an additional solution with no physically new aspects. Variations in the clock problem are introduced to make clear the essential asymmetry which exists between the two clocks.

INTRODUCTION

IN the last several years there has been a revived interest in the problem in relativity known as the clock paradox. This interest has arisen for several reasons. Firstly, with the advent of artificial satellites and planets, space travel is thought of as a reality. Secondly, the development of fantastically accurate atomic clocks has led to plans with satellites for exciting new experiments to test certain relevant predictions of the theory.[1] Thirdly, there has been a lively controversy among some physicists as to what the theory does predict.[2,3]

The problem is usually presented as a question about two identical clocks. Clock A remains on the earth while clock B makes a voyage into space at speeds comparable with that of light and later returns to the earth. Do the clocks record the same time interval for B's space trip? Often twin observers A and B are considered to accompany their clocks, but this is not essential. The accepted answer is that clock B will record a shorter time interval than clock A. The effect is directly connected with the time dilatation of Special Relativity, i.e., a moving clock is observed to run slower than a clock at rest.

The answer to the problem presents a paradox with at least three facets. It is a paradox, in the dictionary meaning of the word, from the viewpoints of (i) absolute time, (ii) the Special Theory of Relativity, and (iii) the General Theory of Relativity.

(i) The layman with an intuitive idea of absolute time is immediately puzzled by any possible difference between the two clocks.

[1] S. F. Singer, Phys. Rev. 104, 11 (1956).
[2] W. H. McCrea, Nature 177, 782 (1956).
[3] E. M. McMillan, Science 127, 158 (1958).

(ii) A physicist is aware of time dilatation based on the Lorentz transformations, but will recall that the effect depends on relative motion and is completely reciprocal, i.e., A finds B's clock runs slow but also B finds that A's clock runs slow. Some may then be puzzled since, although accelerations do occur, the effect is calculated from the velocities alone; because A can be considered as moving relative to B, the problem appears symmetrical in relation to the two clocks and no time difference should occur.

(iii) Other physicists note that there are accelerations and recall that while the Lorentz Transformations of Special Relativity apply to inertial frames of reference, the General Theory of Relativity deals with accelerated frames of reference. They may then assume that the problem can be treated only by the General Theory. If they take such statements of Relativity as "the laws of physics are the same in all frames of reference" too naively, they may be puzzled by the predicted lack of symmetry between the clocks.

THREE SOLUTIONS

Of the many solutions which have been given for the clock problem three approaches which are particularly instructive will be mentioned. Certain results of Special Relativity are assumed: the Lorentz Contraction [a length moving at speed v is observed as reduced by the factor $(1-v^2/c^2)^{\frac{1}{2}}$], the time dilatation [a clock moving at speed v is observed to run slow by the factor $(1-v^2/c^2)^{\frac{1}{2}}$], and the relativistic Doppler factor [for a receding source $[(c-v)/(c+v)]^{\frac{1}{2}}$].

It is convenient to take a simple numerical example and, as is often done, to assume the

Reprinted from Am. J. Phys.
(Nov. 1959 - Pages 580-584)
By Permission

periods of acceleration are negligible compared to the periods of constant velocity. As in Fig. 1, clock A with an observer remains on the earth while clock B with an observer makes a space trip at a speed of $0.6c$ to a star at P, a distance of 6 light-years away. The time taken for the round trip as measured by A is $(2 \times 6/0.6)$ years $= 20$ years. The time dilatation factor for $v = 0.6c$ is 0.8. On the round trip clock B will record $20 \times 0.8 = 16$ years.

(a) Length Contraction

As discussed by Fremlin,[4] the Lorentz Contraction of a moving length can be used to explain the time difference. Once observer B is in motion at speed v, he can measure* the distance AP and will find it not 6 light-years but 6 light-years $\times (1 - v^2/c^2)^{\frac{1}{2}} = 4.8$ light-years. Hence B will calculate his time for the round trip as $(2 \times 4.8/0.6)$ years $= 16$ years. The asymmetry can be thought of as related to the fact that the distance point P is fixed relative to A and not to B.

(b) Doppler Shifts

Darwin, for example, has described how each of the two observers can keep a record of the other's time during the course of the trip by sending out regular time signals.[5] It can be imagined that light or radio signals are sent say every hour in the sender's time. In the example chosen here with $v = 0.6c$ the Doppler factor for the frequency of a receding source is $[(c-v)/(c+v)]^{\frac{1}{2}} = \frac{1}{2}$ and for an approaching source is 2. While B is on the outward trip, A will receive

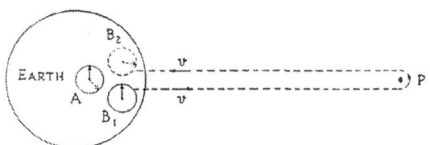

FIG. 1. The clock paradox. Space trip of clock B from earth to star at P and return. Time interval recorded by B is less than that recorded by A.

[4] J. H. Fremlin, Nature 180, 499 (1957).
* The distance can be measured by timing a light signal which may require a considerable lapse of time. Or B can use a range finder, or the angular separation of two light sources at P, and make a correction for aberration.
[5] C. G. Darwin, Nature 180, 976 (1957).

B's signals at 2-hr intervals (corresponding to a red Doppler shift) whereas, when B is approaching, A will receive B's signals at $\frac{1}{2}$-hr intervals (corresponding to a violet Doppler shift).

Consider A's record of B's signal. Observer A will receive slow, i.e., 2-hr signals, from B for the duration of the outward trip *and* for the time that light takes to travel from P to A. For the remainder of the total time A will receive fast or $\frac{1}{2}$-hr signals from B. Hence A will record $(10+6) \times \frac{1}{2} + 4 \times 2 = 16$ years' worth of B's signals.

Then consider B's record of A's signal. B will receive slow signals from A until he reaches P and then, upon reversing his motion, will receive fast signals. If t is the total time of the trip as measured by B then for $t/2$ years B receives slow signals from A and for $t/2$ years fast signals. Since A sends out 20 years' worth of signals during the trip, $(t/2) \times (1/2) + (t/2) \times 2 = 20$ years and therefore $t = 16$ years, as before.

This treatment in terms of Doppler shifts makes it clear that the asymmetry exists because B reverses his motion and hence immediately observes the change in the rate of the signals, whereas A must wait for the time taken by light to traverse the distance P to A before noting the change in the signals.

(c) World Lines

A solution to the problem based on the geometry of space-time is readily appreciated. The element of time τ for clock B is related to the elements of time t and the displacement of B as measured by A in the following way:

$$c^2 d\tau^2 = c^2 dt^2 - (dx^2 + dy^2 + dz^2).$$

Since $d\tau$ must always be less than dt if any motion of B occurs, then for a return trip along any path, which in general will require integration of the foregoing expression, the time interval on clock B will be less than the corresponding interval on clock A. It should be noted that this formulation properly requires that A remain in an inertial frame of reference.

The path of a particle in space-time is often called a world line. A plot of the world lines of the clocks A and B for the example chosen here is given in Fig. 2 where the motion of B is taken to be along the x axis. It is usual to select scales so that the world line of a light signal is at 45°

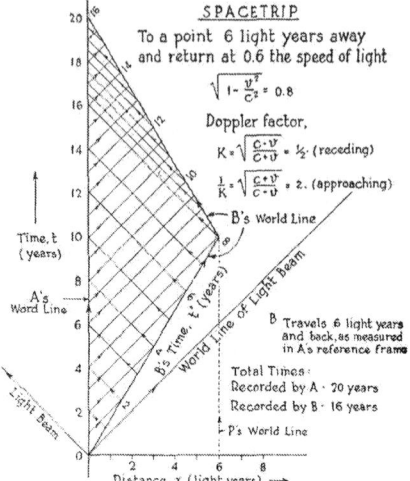

SPACETRIP

To a point 6 light years away and return at 0.6 the speed of light

$$\sqrt{1-\frac{v^2}{c^2}} = 0.8$$

Doppler factor,

$$K = \sqrt{\frac{c-v}{c+v}} = \frac{1}{2} \text{ (receding)}$$

$$\frac{1}{K} = \sqrt{\frac{c+v}{c-v}} = 2 \text{ (approaching)}$$

B's World Line

B Travels 6 light years and back, as measured in A's reference frame

Total Times:
Recorded by A - 20 years
Recorded by B - 16 years

P's World Line

FIG. 2. World lines of clocks A and B showing transmission and reception of light signals.

to the time axis. The units of light-years for distance and years for time give this condition. Clock A, which is at rest in the inertial frame for which the plot has been made, is considered as moving upward along the vertical axis with the flow of time. Clock B moves along its world line, whose slope is proportional to the reciprocal of the velocity, to P and then returns. Typical light signals sent from A and from B at regular time intervals are also shown. The Doppler shifts can be noted in the arrival times of the light signals. The plotted length for units of time on B's world line can be shown to be $(1+v^2/c^2)^{1/2}/(1-v^2/c^2)^{1/2}$ of that for A. For a trip with finite accelerations the world line of B can be modified as shown by the dotted curves. The times would be slightly altered by the necessary integrations for the curved portions of the path.

A plot of A's world line from B's viewpoint can be carried out as done for example by Builder.[6] Since B's frame of reference is accelerated for certain intervals of the trip, peculiarities arise in the plot. Often the necessary calculations are formulated by means of the General Theory. Such a procedure is not essential; the Special

Theory alone can be used as the basis for the calculation.[6]

Since clock B is initially and finally at rest with clock A, the trip must involve accelerations. In the foregoing discussion it was assumed that A remains in an inertial frame and B accelerates. The durations of the accelerations were taken to be negligible compared with the time of the trip. The result that clock B is behind clock A certainly is related to the assumption that B is accelerated whereas A is not, but the difference in the times recorded is calculated from the velocity. Because of their obvious importance, the accelerations require further discussion. The time recorded by clock B during the acceleration can be calculated by an appropriate integration of the form $\int (1-v^2/c^2)^{1/2} dt$ where v is now a function of the time. Bondi has calculated numerical results for trips of different duration in the case where there is acceleration throughout the journey equal to the acceleration of gravity at the earth's surface.[7] The acceleration is reversed at the mid-point of the journey each way. It must be assumed, as is also done in the General Theory, that only the velocity and not the acceleration affects the observed rate of the moving clock. Of course if the acceleration were very great, the rate of a clock might be changed by the attendant inertial force field, but this effect is incidental, just as, for example, temperature effects on the rate of the clock would be incidental.

The clock problem is completely and accurately treated with the use of the Special Theory of Relativity alone and the minor assumption that rate of a clock does not depend on acceleration. There is nothing in the Special Theory which precludes treatment of accelerations; the restriction is that the treatment must be given from the viewpoint of an inertial frame of reference. Once the solution is obtained, all of physical interest is known. A formulation can be set up to yield the same results from a non-inertial frame of reference, but nothing new is thereby added.

USE OF THE GENERAL THEORY

It is often stated or tacitly assumed that the clock paradox can only be resolved by reference

[6] G. Builder, Australian J. Phys. 10, 246 (1957).

[7] H. Bondi, Discovery 18, 505 (1957).

to the General Theory of Relativity. This is certainly so if it is considered that the paradox only arises within General Relativity—the third aspect of the paradox as described above. This aspect of the paradox is by far the most trivial of the three.

The usual resolution of the clock paradox by means of the General Theory is carried out as follows. The first part of the calculation of the times for clocks A and B is made from A's inertial frame of reference using the results of the Special Theory including the necessary integrations for the periods of acceleration. The second part is the calculation from B's frame of reference employing the General Theory more specifically. The results of the two calculations agree. The agreement is then said to have resolved the paradox. In fact, the agreement is merely a verification of the proper formulation of the theory. As has been stressed by Builder,[6] the General Theory adds nothing of physical significance to the problem. The principle of the equivalence of gravitational and inertial mass— the only additional physical basis of the General Theory—is used merely in a formal way and has no direct relevance to the result.

A simple example from classical physics may

elucidate what is done in applying the General Theory to the clock problem. A spiral spring carrying a weight is suspended in a merry-go-round. Given the mass of the weight, the force constant of the spring and the angular speed and radius of the merry-go-round, calculate the extension of the spring. The spring and weight can be viewed from an inertial frame of reference on the earth and a complete solution obtained. Or the problem can be approached from the rotating frame of reference in which case the laws of mechanics are modified by the introduction of a centrifugal force. The same answer will of course be obtained. It is certainly not necessary to use *both* methods to solve the problem. It may be a satisfaction to see that the answers are the same and thereby to verify that the theory applicable to noninertial reference frames has been properly formulated, but essentially nothing new is added by the two solutions.

Textbooks[8,9] and recent articles[10,11] may be misleading in their implication that the clock paradox can be resolved only by the General Theory. What Born and Biem[12] call the "apparent" paradox of the General Theory is indeed resolved, but the General Theory gives little, if any, aid to a physical understanding of the problem.

VARIATIONS OF THE CLOCK PROBLEM

The essential asymmetry which gives rise to the time difference between the clocks can be made clear by several variations of the problem. Figure 3 shows the world lines for the clocks in four cases. The first two cases are where the clocks A and B would record the *same* time interval. In Fig. 3(a), A and B move in opposite directions at the same speed and later reverse their directions and return to the starting point. In Fig. 3(b) B moves off, comes to rest at a distant point, then later A moves off at the same speed and joins B. For the second two cases the clocks would record *different* time

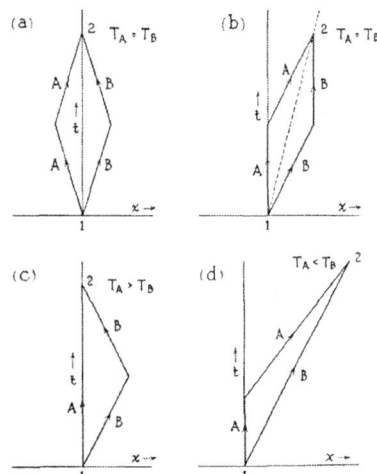

FIG. 3. World line diagrams for variations on the clock paradox problem. In (a) and (b) clocks agree. For (c) B is behind A. For (d) A is behind B.

[8] R. C. Tolman, *Relativity, Thermodynamics, and Cosmology* (Oxford University Press, New York, 1934), p. 194.
[9] C. Møller, *The Theory of Relativity* (Oxford University Press, New York, 1952), pp. 48, 258.
[10] Frye and Brigham, Am. J. Phys. 25, 553 (1957).
[11] Leffert and Donahue, Am. J. Phys. 26, 515 (1958).
[12] Born and Biem, Proc. Koninkl. Ned. Akad. Wetenschap B51, 110 (1958).

intervals. In Fig. 3(c) *B* moves off and later returns to *A*. This is the standard example which has been discussed above. Clock *B* records the shorter time interval. In Fig. 3(d) *B* moves off but instead of stopping maintains the constant speed while *A* after an interval at rest moves off at a higher speed and joins *B*. Here clock *A* will record the shorter time. From these examples it is clear that the clock which changes its velocity when the two are separated will record the shorter time. Changes of velocity, or accelerations, have an absolute significance. In principle they can be observed by "laboratory" experiments, i.e., without direct reference to bodies beyond the observers immediate environment.

Normally the world lines must be plotted from an inertial frame of reference. Replotting of Fig. 3(b) employing an inertial frame corresponding to the dotted world line would make it similar to Fig. 3(a). Likewise replotting Fig. 3(d) using the inertial frame of *B* after the start would produce merely a reflection of Fig. 3(c).

CONCLUSION

The aspect of the clock paradox which is generally found most puzzling is that which arises within Special Relativity. It can be dealt with in a completely satisfactory way by the Special Theory. The usual treatment of the problem by the General Theory does resolve a relatively unimportant aspect of the paradox, but is in fact little more than a verification of the formulation of that theory. The same assumptions made in Special Theory treatment are carried over into the General Theory treatment and no further insight can be gained as to the nature of the effects predicted.

THE

VISUAL

APPEARANCE

OF *RAPIDLY MOVING OBJECTS*

By V. F. Weisskopf

I WOULD like to draw the attention of physicists to a recent paper by James Terrell [1] in which he does away with an old prejudice held by practically all of us. We all believed that, according to special relativity, an object in motion appears to be contracted in the direction of motion by a factor $[1 - (v/c)^2]^{1/2}$. A passenger in a fast space ship, looking out of the window, so it seemed to us, would see spherical objects contracted to ellipsoids. This is definitely not so according to Terrell's considerations, which for the special case of a sphere were also carried out by R. Penrose [2]. The reason is quite simple. When we see or photograph an object, we record light quanta emitted by the object when they arrive simultaneously at the retina or at the photographic film. This implies that these light quanta have *not* been emitted simultaneously by all points of the object. The points further away from the observer have emitted their part of the picture earlier than the closer points. Hence, if the object is in motion, the eye or the photograph gets a "distorted" picture of the object, since the object has been at different locations when different parts of it have emitted the light seen in the picture.

In special relativity, this distortion has the remarkable effect of canceling the Lorentz contraction so that objects appear undistorted but only rotated. This is exactly true only for objects which subtend a small solid angle.

In order to understand the situation thoroughly let us consider the distortion of the picture we see of a moving object under nonrelativistic conditions, where light moves with light velocity c only in the stationary frame of reference of the observer, and a moving object does not suffer a Lorentz contraction. In the frame of the object moving with the velocity v the light velocity would be $c - v$ in the direction of motion and $c + v$ in the opposite direction.

We first consider the case of a cube of dimension l moving parallel to an edge and observed from a direction perpendicular to the motion. The observation is made at great distance in order to keep the subtended angle small (see Fig. 1). The square $ABCD$ facing the observer will be seen undistorted since all points have the same distance from the observer. The square $ABEF$ facing in the opposite direction of the motion (the rear side in regard to the motion, not in regard to the observer's position) is invisible when the cube is not in motion. However when it moves it becomes visible since the light from E and F is emitted l/c seconds earlier when the points E and F were $(v/c)l$ further behind at E' and F'. Hence the face $ABEF$ will be seen as a rectangle with a height l and a width $(v/c)l$. The picture of the cube, therefore, is a distorted one. In an undistorted picture of a rotated

[1] J. Terrell, Phys. Rev. **116**, 1041 (1959).
[2] R. Penrose, Proc. Cambridge Phil. Soc. **55**, 137 (1959); see also H. Salecker and E. Wigner, Phys. Rev. **109**, 571 (1958).

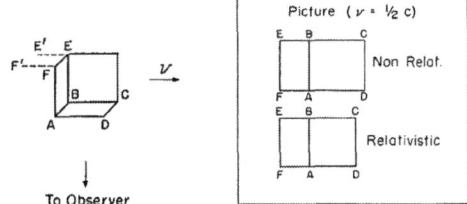

Fig. 1. A cube moving with velocity v seen by an observer at an angle of 90°.

Reprinted from Phys. Today
(Sept. 1960 - Pages 24-27)
By Permission

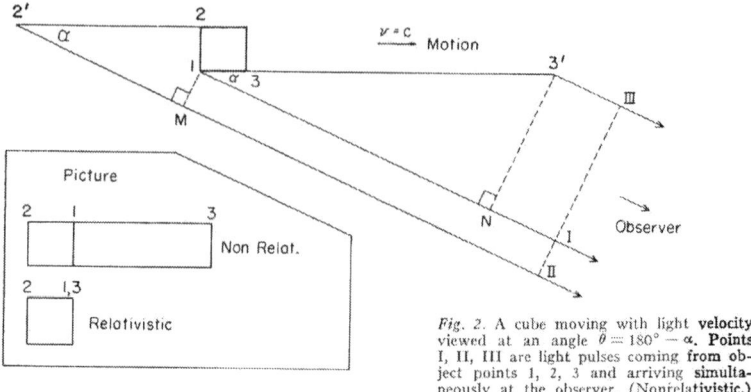

 placed above; figure content described by caption:

Fig. 2. A cube moving with light **velocity** viewed at an angle $\theta = 180° - \alpha$. **Points** I, II, III are light pulses coming **from** object points 1, 2, 3 and arriving **simultaneously** at the observer. (Nonrelativistic.)

cube both faces should be foreshortened; if the face *ABEF* is shortened by the factor v/c, the other face *ABCD* should be foreshortened by $(1 - v^2/c^2)^{1/2}$, whereas here *ABCD* appears as a square. Hence the picture of the cube appears dilated in the direction of motion. A similar consideration for a moving sphere shows that it would appear as an ellipse elongated in the direction of motion by a factor $(1 + v^2/c^2)^{1/2}$.

We get even more paradoxical results by considering the picture of a moving cube in a nonrelativistic world, seen not at 90° to the direction of motion but at $180 - \alpha$ degrees where α is a small angle. We now look at the object to the left when it is coming towards us from the left. We will assume now that $v/c = 1$ in order to simplify our considerations. What is the picture then? Fig. 2 illustrates the situation. The edges *AB*, *CD*, *EF* are denoted by the numbers 1, 2, 3. We assume that the edge 1 emits its light quanta at the time $t = 0$. Where must the edges 2 and 3 emit their light such that it travels in a common front with the light from 1, in order to arrive simultaneously at the observer? It is easily seen that 2 must emit its light much earlier; in fact it must happen when it is at the point marked 2′ which is determined by the equality of the distances (2′2) and (2′M). The interval (2′2) is the distance which the edge 2 travels between the emission of light by 2 and 1. The length (2′M) is the distance which the light travels from 2′ in order to be "in line" with the light emitted by 1. Both light and edge travel with the speed c. We can see that the distance (1M) is equal to (1 2) which is the size l of the cube. The light seen from edge 3 is emitted much later, when

Victor F. Weisskopf is a member of the Laboratory for Nuclear Science and professor of physics at the Massachusetts Institute of Technology, Cambridge, Mass.

the edge is at 3′. The point 3′ is determined by the equality of the distances (3 3′) and (1N). A simple calculation shows that $(3'N) = l \sin \alpha \, (1 - \cos \alpha)^{-1}$.

What then is the picture we see of the cube? It is indicated in the figure by the points I, II, III which represent the positions of the light quanta coming from the object and form the picture. We will see a strongly deformed cube with the edge 1 in the middle, the edge 2 on the left of 1 as if we were looking from behind (from the left to the right) and the edge 3 quite far to the right of 1. Again we see a picture elongated in the direction of flight. The face between the edges 1 and 2 appears as a true square.

We now will show that relativity theory simplifies the situation. It removes the distortion of the picture and what remains is an undistorted but rotated aspect of the object. We can see this directly with the examples quoted. Consider the cube when looked at perpendicular to its motion; the Lorentz contraction reduces distance between the edges *AB* and *CD* by the factor $(1 - v^2/c^2)^{1/2}$ and leaves the distance between *AB* and *EF* unchanged. Therefore the picture of the face *ABCD* is foreshortened precisely by the amount necessary to represent an undistorted view of a cube turned by an angle whose sine is v/c. In the case of the cube moving with light velocity towards us the Lorentz contraction reduces the distance between the edges 1 and 3 to zero. The picture one sees then is a regular square corresponding to the rear face and nothing else, since edge 3 coincides with edge 1. Hence we see an undistorted picture directly from behind. The object is undistorted but turned by an angle of $(180 - \alpha)$ degrees.

We can show by means of the following consideration that this result is quite generally true for any object. Let us consider an assembly of light pulses originating from N points of the object, traveling all

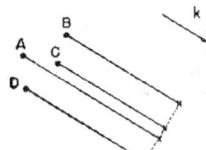

Fig. 3. A "picture". A, B, C, D are points of the object. The four crosses are the light pulses making up the "picture".

in the same direction given by a vector **k**, and such that the light pulses are all in one plane perpendicular to **k** (see Fig. 3). Then they will arrive simultaneously at the eye of the observer and produce the shape which is seen. We will call such an assembly of light pulses a "picture" of the object. Under nonrelativistic conditions a "picture" does not remain a picture when seen from a moving frame of reference. The reason is that, in the moving frame, the plane of the light pulses is no longer perpendicular to the direction of the propagation. In a relativistic world a "picture" remains a "picture" in any frame of reference. The light pulses would arrive simultaneously at a camera in every system of reference.

This fact can be proven immediately in the following way: The light pulses form a wave front or can be imagined as moving embedded in an electromagnetic wave exactly where this wave has a crest. It is known that electromagnetic waves are transverse in all frames of reference. That means that a wave front or the plane of the wave crest is perpendicular to the direction of propagation in *any* system. We can also show that the distance between the light pulses is an invariant magnitude. Here we only need to introduce a coordinate system where the x-axis is parallel to the propagation. Then for two light pulses of the picture the invariant $(x_1 - x_2)^2 + (y_1 - y_2)^2 + (z_1 - z_2)^2 - c^2(t_1 - t_2)^2$ is equal to the square of the distance d between the two pulses, since $d^2 = (y_1 - y_2)^2 + (z_1 - z_2)^2$ and $x_1 = x_2$ when $t_1 = t_2$. The latter relation expresses the fact that the pulses are in a plane perpendicular to the propagation.

The only thing that is not invariant is the direction of propagation, the vector **k**. The transformation of this direction is given by the well-known aberration formula. A light beam whose direction includes the angle θ with the x-axis is seen including an angle θ' with the x-axis in a system moving with the velocity v along the x-axis: *

$$\sin \theta' = \frac{(1 - v^2/c^2)^{\frac{1}{2}} \sin \theta}{1 + (v/c) \cos \theta}.$$

We can conclude the following result from the invariance of the "picture": The picture seen from a moving object observed at the angle θ is the *same* as one would see in the system where the object is at rest, but ob-

* The angles refer to the direction in which the light beam is seen; that means a direction opposite to the motion of the light pulses.

served at the angle θ'. Hence we see an undistorted picture of a moving object, but a picture in which the object is seemingly rotated by the angle $\theta' - \theta$. A spherical object still appears as a sphere.

This must not by any means be interpreted as indicating that there is no Lorentz contraction. Of course, there is Lorentz contraction, but it just compensates for the elongation of the picture caused by the finite propagation of light.

It is instructive to plot the angle θ' as a function θ. Fig. 4 shows this relation for $v = 0$, for a small

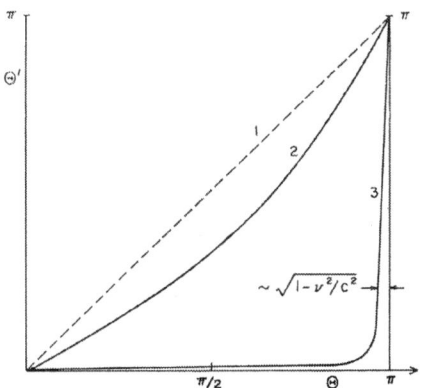

Fig. 4. The angle of observation θ' of a light beam relative to the direction of v, seen by an observer in the moving system, versus the same angle θ as seen in the rest system. Curve 1 is for $v = 0$, curve 2 is for $v = c/2$, curve 3 is for $v = c$.

value of v/c and also for the case $v/c \approx 1$. We see that the apparent rotation is always negative, which means that the object is turned such that it reveals more of its trailing side to the observer. In the extreme case of $v \approx c$, θ' is extremely small for all values of θ except when $180 - \theta$ is of the order $[1 - (v/c)^2]^{1/2}$. Since θ goes from 180° to 0° when an object moves by, we find for the case $v \approx c$ that we see the front side of the object only at the very beginning; it turns around facing its trailing side at us quite early when we still see it coming at us and remains doing so until it leaves us and naturally is seen from behind. This paradoxical situation is perhaps not so surprising when one is reminded of the fact that the aberration angle is almost 180° when $v \approx c$. Hence the light which we see coming from the object when it is moving towards us, has left the object backwards when observed from the object itself.

The situation becomes clearer when we look closer at the distribution of the emitted light as seen from the observer. Let us assume that the moving object emits radiation which is isotropic in its own rest system, i.e., its intensity is independent of the emission angle θ'. This radiation does not at all appear isotropic in the nonmoving system; there it seems concentrated in the forward direction. If $v \approx c$, most of the light appears to be emitted such that it includes a very small angle θ with v. This is a well-known effect which causes an isotropic emission to look as if almost all radiation is emitted in the form of a focused headlight beam. One example of this effect is the radiation of electrons running along a circle with a velocity near c as one finds it in synchrotrons. In this case the radiation in the rest system is not completely isotropic; it is essentially a dipole radiation. Still it appears as an emission sharply peaked in the direction of flight.

The apparent angular distribution $I(\theta)$ of the radiated intensity in the system at rest is connected with the angular distribution $I_0(\theta')$ in the system of the moving object by an expression which is related to the aberration formula:

$$\frac{I(\theta)}{I_0(\theta')} = \frac{\sin \theta' d\theta'}{\sin \theta d\theta} = K(\theta)$$

$$K(\theta) = \frac{1 - (v/c)^2}{[1 + (v/c) \cos \theta]^2}$$

where θ is the angle of observation, which means that the forward direction is near $\theta = \pi$. $K(\theta)$ is plotted in Fig. 5 as a function of θ and we see that the width of the "headlight" beam is of the order $[1 - (v/c)^2]^{1/2}$.

The factor $K(\theta)$ also determines the Doppler shift of the light. If the emitted light has the frequency ω_0 in the system moving with the object, the observer at rest sees a frequency $\omega = K^{1/2}\omega_0$. The frequency is increased or decreased by the square root of the factor by which the intensity is enhanced or reduced.

We note that, for $\theta = 90°$ there is a reduction of frequency due to the relativistic Doppler shift.

We now describe what is seen when an object is moving by with a velocity near that of light. First, when the angle of vision is still near 180°, we see the front face of the object, strongly Doppler-shifted to very high frequencies and with high intensity. We are looking into the "headlight" beam of the radiation. When the angle of vision becomes of the order $\pi - [1 - (v/c)^2]^{1/2}$ the color shifts towards lower frequencies, the intensity drops, and the object seems to turn. When $\theta \approx \pi - 2^{1/2}[1 - (v/c)^2]^{1/4}$, still an angle close to 180°, the intensity becomes very low, we are out of the "headlight" beam, the color is now of much lower frequencies than it would be in the system moving with the object; the object has turned all around and we are looking at its trailing face. The front is invisible because the beams emitted forward in the moving system are concentrated into the small angle of the "headlight". The picture seen at angles smaller than $\pi - 2^{1/2}[1 - (v/c)^2]^{1/4}$ remains essentially the same until the object disappears. It is the picture expected when the object is receding. However it appears already when the object is moving toward us.

We would like to emphasize that all these considerations are exact only for objects which subtend a very small solid angle. Only then the picture consists essentially of parallel light pulses. If the angle subtended is finite we must expect different rotation angles for different parts of the picture and this would lead to some distortions. It has been shown by Penrose,[2] however, that the picture of a sphere retains a circular circumference even for large angles of vision.

It is most remarkable that these simple and important facts of the relativistic appearance of objects have not been noticed for 55 years until J. Terrell discovered and fully recognized them in his recent publication.

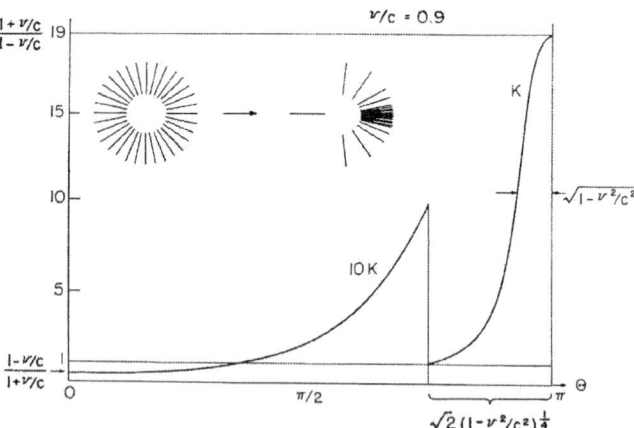

Fig. 5. The ratio K of the emitted intensity per solid angle measured in the observer's system to the intensity per solid angle measured in the system moving with the object, as function of the angle θ of observation. K also determines the Doppler shift. The observed frequency ω is related to the emitted frequency by $\omega = K^{1/2}\omega_0$.

Time and Relativity: Part I.

by O. R. FRISCH

Cavendish Laboratory, Cambridge

The questions I shall discuss are not new, nor are the answers: they can mostly be found in Einstein's early papers (1905, 1911). But the coming of artificial earth satellites has offered new possibilities for testing some of Einstein's predictions, and some tests have actually been carried out with the help of the recoilless gamma radiation (Mössbauer 1959). Furthermore Dingle (1956, etc.; for a review see Sherwin 1960) has cast doubt on some consequences of Einstein's theories, and the ensuing controversy has left many people thoroughly bewildered. Hence a presentation of the behaviour of time according to the theory of relativity might be useful.

Time is of course measured with clocks, and there is a variety of them. A grandfather clock would clearly be no use in a spaceship where gravity may be absent (or greatly increased as during take-off), and even a wristwatch may be slightly affected by acceleration. But that is no essential difficulty: an atomic clock—which uses the frequency associated with some atomic transition— would be far less sensitive to acceleration, and nuclear frequencies are less sensitive still (Sherwin 1960).

To appreciate what the relativity theory has to say about time, we recall what Newton said in his Principia: " Absolute, true, and mathematical time, of itself and from its own nature, flows equably without relation to anything external . . .". That assumption has been criticized because it cannot be tested: Newtonian time is ticking away all through space, as it were, but its ticks cannot be heard. However, we must remember that Newtonian physics has no speed limit; so the signal from a master clock can be sent to anywhere in space with as little delay as you please so that you can tell what time it is at any place without ambiguity.

Relativity—in a limited sense—is part of Newtonian physics. A frame of reference relative to which any free mass point moves in a straight line at constant speed is called an inertial frame, and it was known at Newton's time that any frame that moves at constant speed in a straight line relative to an inertial frame is again inertial. That fact, that the laws of mechanics are equally valid in any two inertial frames is sometimes called the Galilean principle of relativity. It does not embrace the propagation of light (or other electromagnetic phenomena). But from about 1887 (Michelson and Morley) evidence accumulated that electromagnetism ought to be included in the principle of relativity; in particular, light appeared to have the same speed c relative to all frames of reference. Various attempts were made to modify mechanics accordingly; but the first simple and consistent presentation was given by Einstein in 1905.

In this article I shall not try to present all of relativity theory, but only those features which relate to the question: How do moving clocks behave? The most striking features of that behaviour are these:

(1) Two events, simultaneous by the clocks in one inertial frame, will in general not be simultaneous by the clocks in another inertial frame.

(2) of two identical clocks in relative motion, each will be observed to go slow by an observer moving with the other.

Reprinted from Contemporary Phys.
(Oct. 1961 - Pages 16- 27)
(Feb. 1962 - Pages 194-201)
By Permission

(3) of two identical clocks in a gravitational field, the one at higher gravitational potential will go faster.

These features I shall derive as directly as possible from these two basic assumptions: (1) any two inertial frames are equivalent, and (2) light always travels at the same speed. In fact it is best to combine these two into one sentence which expresses the special relativity principle:

Any two inertial frames are equivalent, also with regard to the propagation of light.

From this one assumption we can derive the statements (1) and (2) about clocks. To derive (3) we also need the principle of the universal proportionality of weight and mass (also called the equivalence principle); this will be discussed in a second article.

The decisive difference from Newtonian physics is that we now have a speed limit: no signal can travel faster than light. So to synchronize two clocks we must make do with messengers that have finite speed. In principle, all the subsequent arguments could be carried through if we employed boys on bicycles to set the clocks, but we would have to know how their speed transforms from one frame of reference to another. So it is better to use light as a messenger; its constant speed (in vacuo) relative to all frames of reference makes all arguments much simpler.

Let us start with a simple problem: how would Jack and Mac, several thousand million miles apart, but both at rest in the same inertial frame S, synchronize their clocks? Mac might start by adjusting his clock so that it always reads the same time that he can see—through his telescope—on Jack's clock. But soon he gets a message that his clock, as seen by Jack, is several hours behind; so he splits the difference by advancing his clock by half that amount. Now both clocks behave in the same way: each, as seen from the other clock, appears late by the same amount (the time the light takes from one to the other). Hence they can now be considered synchronized according to the relativity principle.

An alternative way of synchronizing those clocks would have been to set them both by a master clock, placed half-way between Jack and Mac. The symmetry of that procedure guarantees in a transparent way the equivalence of the two clocks after setting, and the result is of course the same.

Let us now place a coordinate system in S so that Jack is in its origin, and Mac on the positive x-axis. Next we introduce two more clocks, belonging to Ed and Fred who travel along the x-axis at the speed v, with Fred in front. Since everything happens on the x-axis we can ignore the y and z axes and draw the usual space-time graph, fig. 1. A point on that graph means an event, and we shall look at the two events O (Ed passes Jack) and P (Fred passes Mac). By chance O and P occur at the same time, according to the clocks in S, i.e. Jack's and Mac's. But not so according to Ed and Fred (frame S'): They will have synchronized their clocks by the same procedure, say, with a master clock that moves along, half-way between them, and a glance at fig. 1 shows that by their clocks the event synchronous with O is not P but Q, which is later. Both pairs of clocks have been synchronized by the relativity principle. If one pair, say Jack and Mac, is considered at rest, then of the two moving clocks the one in front (in space) is behind (in time). Of course the relativity principle

C.P.   B

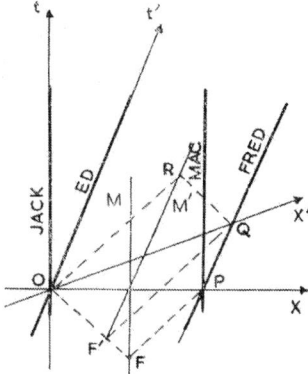

Fig. 1. Relativity of Simultaneity. In this space-time diagram, each point represents an
event. Events at the same place in the inertial system S (the 'rest frame') lie on
a line parallel to the t-axis (chosen vertical); simultaneous events lie on a line parallel to
the x-axis (chosen horizontal). The travel of a light signal is shown by a dotted line.

Jack and Mac are represented by bold vertical lines since they are both at rest in
S. M is a master clock half-way between them; a light flash emitted at F will
reach them at O and P; for reasons of symmetry O and P must be simultaneous
in S, confirming our choice of the horizontal direction for the x-axis.

But in S', the moving inertial system in which Ed and Fred are at rest, O and P
are not simultaneous. A flash emitted at F' from the master clock M' (at rest in S'
and half-way between Ed and Fred) will reach Ed and Fred at O and Q respectively;
so O and Q are simultaneous in S' and Q is later than P. Fred's clock will read
less as he passes Mac than Ed's clock reads as he passes Jack although Jack and Mac
record those two events as simultaneous: "The clock in front is behind in time".

If we choose the origin of S' (i.e. the point $x'=0$, $t'=0$) also at O, then the
x'-axis must connect O with Q since O and Q are simultaneous in S'. It is convenient to choose length and time units so that the speed of light $c=1$; then the
'light lines' are at 45° to the vertical and with the help of the two auxiliary lines
OR and QR one sees easily that $\sphericalangle xx' = -\sphericalangle tt'$, a fact that is sometimes useful to
remember.

allows us to consider S' at rest and S in motion, in the opposite direction: then
Jack is ahead of Mac in space and indeed again behind in time.

It is surprisingly easy to get wrong results by overlooking that 'relativity
of synchronism'. For instance take the following arguments: the light energy
from the flare-up of a nova is contained in an expanding spherical shell, with the
nova at the centre; a body, flung off at great speed by the explosion, would not
be at the centre of that shell; 'hence relative to that body the light has not
been spreading at the same speed in all directions'. That conclusion contradicts
the relativity principle and must be wrong. The fallacy is that the shell we
mentioned contains the light energy at a given *time relative to the nova*; if we
ask where the light energy is at a given *time relative to the moving body* we shall
again find a spherical shell, now with that body at the centre. The motion
of the nova when it exploded is quite irrelevant; relative to any inertial frame
the shell is centered about the point where, *in that frame*, the nova was when
it exploded.

By the way, the expanding ring around the Nova Persei of 1901 is not that shell. The shell as such can never be seen: the light from its different parts would reach us at widely different times. What we see today are those scattering particles by which the light can reach us with 60 years' delay: they form an ellipsoidal shell with us at one focus and the place (in our inertial frame) of the nova explosion at the other. That light reaches us from all directions; but it looks weak near the centre because back-scattering is weak, and falls off again with increasing distance from the explosion, giving the observed pattern of a diffuse ring.

Now to our second statement, concerning the behaviour of clocks in relative motion. Let us see how Jack and Ed would synchronize their clocks. Setting them is easy: as they pass each other they will set their clocks to the same time, say zero. But after that they will move apart and light signals will take more and more time from one to the other. Let us assume that Ed at first adjusts the rate of his clock by the receding face of Jack's clock, ignoring the growing delay. His clock then reads unit time when he sees unit time on Jack's clock; but Jack reports that he had to wait the longer time $(c+v)/(c-v)$—see fig. 2— before he saw unit time on Ed's clock. Once again one has to split the difference, or rather the ratio: Ed speeds up his clock by the factor $f = \sqrt{\{(c+v)/(c-v)\}}$ and resets it so that it extrapolates back to zero as before. Now the clocks are again equivalent: each clock keeper sees the receding clock go slower by the factor $1/f$.

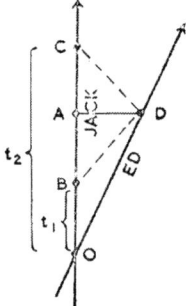

Fig. 2. Relativistic Doppler effect. Both Jack and Ed have set their clocks to zero when they passed each other (event O). Some time later (event D) Ed observes that Jack's clock reads t_1 (event B) and naïvely he sets his clock to t_1 as well; but soon Jack reports that he had to wait until the time t_2 before (event C) he saw t_1 on Ed's clock. If the units are chosen so that $c = 1$, then $AB = AC = AD = (v/c) \cdot AO$; it follows that $t_2/t_1 = OC/OB = (c+v)/(c-v)$. So Ed should have adjusted his clock to a speed $f = \sqrt{\{(c+v)/(c-v)\}}$ times higher than he did; when that is done the two clocks are equivalent in that each appears slow by the factor $1/f$ if viewed from the other.

Again, the synchronization could have been done more simply by the use of a master clock halfway in between. In order to guarantee the symmetry of the set-up we choose our rest system so that the master clock is at rest at its origin, and we assume Tim and Jim with their two clocks move away from it at speeds $+v$ and $-v$ respectively. Once we have done that we can forget the

O. R. Frisch

master clock: clearly the two clocks will be synchronized if their 'ticks' are represented by equal lengths in fig. 3. When Tim looks back at Jim's clock it will again read less than his own, and the ratio of the two readings is found (see fig. 3) to be $(c+v)/(c-v)=f^2$. That agrees with the result obtained in the last paragraph: Tim sees the master clock go slow by the factor $1/f$ and the keeper of the master clock sees Jim's clock go slow by that same factor; the two factors must be the same because the relative velocities are the same both times. (By the way, the relative speed of Tim and Jim is not $2v$ but $2v/(1+v^2/c^2)=V$, and the factor f replaced by $F=\sqrt{\{(c+V)/(c-V)\}}=(c+v)/(c-v)=f^2$.)

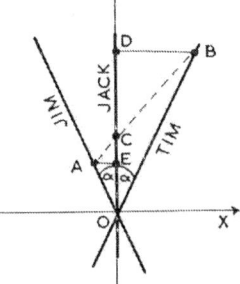

Fig. 3. Another derivation of the Doppler factor f. Tim and Jim are travelling away from Jack in opposite directions, at speed $v=c.\tan\alpha$. In Newtonian physics Jim's clock would appear slow to Jack by the factor $EO/CO=1/(1+v/c)$, whereas Jack's clock appears slow to Tim by the factor $CO/DO=1-v/c$; the Doppler factor here depends on whether the source or the receiver of the signals is at absolute rest. In relativity theory the two factors must be alike, but their product must be $(c-v)/(c+v)$ as before; hence the factor must be $\sqrt{\{(c-v)/(c+v)\}}=1/f$, as derived in Fig. 2.

This of course is nothing but the Doppler effect. We have spoken of discrete time signals sent from one clock to the other, but it would be equally true for the successive crests of a long radio wave where the wave crests can be observed on an oscillograph; and since there is no difference in principle between radio waves, light or gamma rays, we conclude that our factor $1/f$ applies equally to the frequencies of all those radiations. Light from a receding source will be red-shifted, its wavelength increased by the factor f; if instead the source approaches we get a blue-shift, and by replacing v by $-v$ in the expression for f we find that the wavelength is now shortened by the factor $1/f$. Any source of monochromatic radiation is a clock for our purposes, though with light or gamma rays we cannot count the individual ticks.

Of course the Doppler effect was known in classical physics, but there it depended on the absolute motion of the clocks (i.e. their motion relative to the ether). Figure 3 shows that the frequency of the receding clock was decreased by the factor $(1-v/c)$ as seen from the clock at rest whereas the other way round the factor was $1/(1+v/c)$. The relativistic factor $1/f$ is seen to be the geometric mean between those two classical values. By the way, the classical formulae can still be applied, e.g. to the Doppler effect of sound in still air, if the sound velocity is called " c ".

Let us now go back to relativity and ask for the time recorded by Ed's clock while unit time passes in the rest system. The simplest way to do that is by using two clocks at rest, say, those of Jack and Mac. Let us assume they are the distance v apart; then if Ed passes Jack at time zero he passes Mac at time 1 (by Mac's clock). Jack watches for that event and sees it when his own clock reads $(1+v/\mathbf{c})$. Ed's clock seems to him slow by the factor $1/f$, so it reads $(1+v/\mathbf{c})/f = \sqrt{\{1-(v/\mathbf{c})^2\}}$ at the time it passes Jim, whose clock, as we saw, reads unit time. So here is our answer: judged by two synchronized clocks at rest, a clock moving at the speed v goes slow by the factor $\sqrt{\{1-(v/\mathbf{c})^2\}}$. That is the reciprocal of the 'Lorentz factor' $\gamma = (1-(v/\mathbf{c})^2)^{-\frac{1}{2}}$, the ratio of the relativistic mass and the rest mass of a body moving at speed v; $1/\gamma$ gives the 'Lorentz contraction', the factor by which a rod appears shortened when judged from a frame in which it moves (lengthwise) at speed v.

This is the 'relativistic time dilatation'. It is not in conflict with the relativity principle; I have been careful to say 'judged by *two* synchronized clocks . . .'. To time a clock moving at uniform speed relative to an inertial frame we need two (or more) synchronized clocks in that frame. Thus we compare one clock with two others, a situation which is essentially unsymmetrical. We can of course ask about the rate of Jack's 'resting' clock with respect to the two clocks of Ed and Fred; relative to them, Jack's clock will now be losing, in complete agreement with the relativity principle.

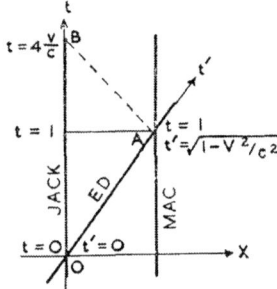

Fig. 4. Relativistic time dilatation. Let Jack and Mac be the distance v apart where v is Ed's speed; then Ed, having passed Jack at zero time, reaches Mac when Mac's clock reads unit time. That event is seen by Jack at the time $1+v/\mathbf{c}$ (event B). The reading on Ed's clock must be $(1+v/\mathbf{c})/f$ where $f = \sqrt{\{(\mathbf{c}+v)/(\mathbf{c}-v)\}}$ is the Doppler factor. So Ed's clock reads $(1-v^2/\mathbf{c}^2)^{\frac{1}{2}} = 1/\gamma$ as the time difference between the two events O and A whereas that time difference is one time unit according to the clocks of Jack and Mac. Hence a moving clock appears slow by the factor $1/\gamma$ when timed on passing two synchronized clocks at rest.

It is vague and misleading to say 'a moving clock goes slow'. To be precise, one should say: 'a clock moving at speed v relative to an inertial frame containing synchronized clocks is found to go slow by the factor $1/\gamma = \{1-(v/\mathbf{c})^2\}^{\frac{1}{2}}$ when timed by those clocks'. That statement does not contradict the relativity principle but indeed follows from it.

O. R. Frisch

There is a way of comparing a moving clock with just one clock at rest, if the moving clock changes speed and comes back to pass the other clock a second time. Let us say Albert with his clock accompanies Ed till they reach Mac; at that point Albert reverses speed and returns to Jack, taking the same time as for the outward journey. Jack will say to him ' you have been away two time units ', but Albert's clock will read only $2/\gamma$. This really seems to mean that a moving clock goes slower than one at rest, and that would constitute a paradox: from the relativity principle we seem to have derived a result that contradicts it. But again our formulation was inaccurate: Albert's clock has lost time relative to Jack's, not because it has moved but because it has changed its motion on the way. So the situation once again is unsymmetrical.

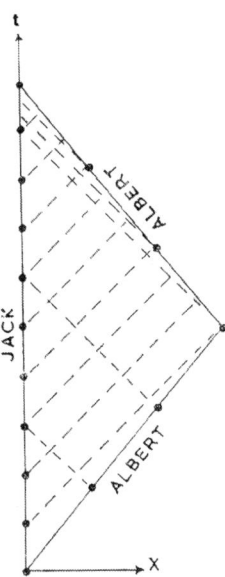

Fig. 5. An example of the ' Twin Paradox '. Albert leaves his twin brother Jack, going for three years (by his clocks) at the speed $0.8\ c$; then he reverses speed and returns home after another three years. His Christmas greetings at first are delayed by the Doppler factor $(1-0.8)/(1-0.8)=3$; but after nine years Jack sees Albert approach, his greetings come every four months, and at the end of the tenth year he is home. He in turn received only one of Jack's greetings at the very end of his outward journey (one every three years, the same Doppler factor), and nine on the way back (three a year). It all fits the expected time dilatation: $1/\gamma = \sqrt{\{1-(0.8)^2\}} = 6/10$.

It may help to discuss an example, designed to give simple numerical relations (Darwin 1957). Albert leaves his twin brother Jack—who is said to be at rest—at X-mas at a speed of 0.8 c and travels for three years, by his own clocks; he then reverses speed and gets home after another three of his years; so he would say, he has been away for six years in all. Both Jack and Albert send each other regular X-mas greetings by radio. The Doppler factor is $\sqrt{\{(1+0.8)/(1-0.8)\}} = 3$; hence Albert receives the first message only after three years, just as he turns back. But during the return journey he gets three messages a year, the last one just as he gets home. So he has received one message on the outward journey (at the end of it), and nine on the way back; hence ten years must have passed for Jack while Albert was away six of his years. This fits: the factor $1/\gamma = \sqrt{(1-16/25)} = 3/5$.

How does this exchange of messages look from Jack's point of view? He receives three messages from Albert on the way out, arriving at intervals of three years, because of the Doppler factor. So he sees Albert receding for nine years. But then the signals speed up: three arrive within one year, and with the last one, Albert is home, after ten years, although he has sent only six annual messages. It all fits; both of them observe the correct Doppler slowing (or speeding) of the other's clocks—through the annual messages—depending on whether they see the other one receding or approaching. But to Jack the outward journey of Albert appears nine times as long as his return trip. So the different times recorded by the two are fully consistent with the relativity principle; any other result would indicate that their clocks had not been synchronized according to that principle.

The difference between their clocks is that Jack's has been at rest in the same inertial system all the time while Albert's has been at rest in two different inertial systems. The latter must therefore have suffered acceleration, during its change of speed, and it has been sometimes suggested that time is lost during that acceleration. But actually those accelerations can easily be kept so small that they would hardly affect a wristwatch, let alone an atomic clock. Furthermore, if the overall journey was made longer the time lost would go up in proportion with the duration of the journey, while any effect of acceleration would remain unchanged and hence become less relevant.

Let me point to a simple analogy. Two motorists go from A to C, Joe in a straight line, Bill via B. On arrival they find that Joe has travelled 60 miles and Bill 70. Surely this is no paradox! Admittedly Joe has travelled in a straight line all the time and Bill most of the time, but no one would say that Bill acquired his extra mileage at the corner he had to turn at B. We are all familiar with the fact that a broken line is *longer* than a straight line between the same two points. But most of us are not yet familiar with the fact—which follows from the relativity principle—that the time interval between the same two events is *shorter* when measured along a broken line (i.e. by a clock that changes its speed) than when it is measured along a straight line (by a clock travelling at constant speed).

Sometimes the objection is made that in general relativity all frames of reference are equally admissible and that therefore the situation between Jack and Albert is not really unsymmetrical; that argument will be discussed in a second article.

Some people are willing to believe that clocks behave like that, but doubt whether Albert would really return looking four years younger than his stay-at-home twin. That is indeed doubtful; he may well have aged a lot more than Jack, because of the discomforts of space travel! But an organism is a clock, though a poor one, easily affected by its surroundings; if those extraneous effects are eliminated or allowed for it must behave like any other clock if the relativity principle is valid.

There is, however, yet another way for comparing two clocks in relative motion, if we bring in another co-ordinate of space. Let us say, Jack wishes to compare his clock with that of the distant traveller Dan, who moves at speed v parallel to the x-axis and at distance b away from it. If b is large enough Dan will spend an appreciable time near enough to the y-axis (fig. 6) so that the signals he sends out during that time will all suffer practically the same delay in reaching Jack; then Jack can just look at Dan's clock, without any need for corrections. Now Dan is at rest relative to Ed, and we remember that Ed's clock is slow by the factor $\gamma = \sqrt{\{1 - (v/\mathbf{c})^2\}}$ if timed by Jack and Mac as it passes them; so we conclude that Dan's clock must be slow by the same factor if timed in the same way. But actually that proviso is unnecessary: Dan's signals are received by Jack and Mac essentially simultaneously, provided they come from a distant point near the y-axis. So the timing may be done entirely by Jack, with the same result: that Dan's clock goes slow by the usual factor $1/\gamma$.

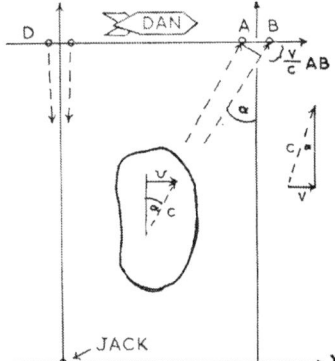

Fig. 6. The 'transverse Doppler effect'. This is *not* a space-time diagram. The bold arrow DB indicates the path of the distant traveller Dan, the two dotted lines on the left the paths of two light signals, sent out one time unit apart by Dan while close to the y-axis. They will travel the same distance (in first order); Jack will see them from the y-direction and receive them $1/\gamma$ time units apart, because of time dilatation (see the text for the detailed argument). What about time signals from Jack, seen by Dan from the y direction? They do not travel parallel to the y-direction in the rest frame but at an angle α where $\sin \alpha = v/\mathbf{c}$ (see insert). The time interval between their arrivals at Dan's at the points A and B is AB/v (since Dan travels at the speed $v = 1 + (v/\mathbf{c}) \cdot$ AB) or equal to $1/(1 - v^2/\mathbf{c}^2)$. This is longer by the expected factor γ than the time unit on Dan's clock; so Dan, looking along the y-axis, sees Jack's clock slow by the factor $1/\gamma$ just as Jack sees Dan's clock.

But is this not another paradox? Here we have two clocks, each at rest in an inertial frame, synchronized according to the relativity principle; how then can one of them go slower than the other?

Let us be precise. What Jack observed is that Dan's clock signals, received from a direction perpendicular to the x-axis (their line of relative motion) are slow compared to his own. The relativity principle demands that the reverse statement, obtained by exchanging the names 'Jack' and 'Dan', should also be true. And that is indeed so. The point is that the signals that Dan sees arriving from a direction parallel to the y-axis do not travel in that direction *relative to the rest frame*: they travel at the angle α relative to the y-axis where $\sin \alpha = v/c$. So in the rest system the delay between two light signals in reaching Dan is not the same although it is (practically) the same if referred to Dan's system. Let us compute that difference in delay, referring everything to the rest system. Let A and B be the two points at which two light signals, emitted one time unit apart by Jack's clock, reach Dan. The second light signal has the extra distance (v/c) . AB to go, so they arrive $(1+(v/c))$. AB time units apart, and in that time Dan has covered the distance AB, going at the speed v. Hence $(1+(v/c))$. AB $= $ AB$/v$, or AB$/v = 1/(1-(v/c)^2)$. That is the interval between the arrivals of the two unit ticks from Jack's clock, and it is indeed longer than Dan's time unit, by the expected factor γ; so he will report that Jack's clock goes slow by that factor, as we know he must.

I hope I have made it clear that the behaviour of clocks is not really paradoxical in the cases I have discussed. This does not prove that no real paradox, no logical inconsistency, can be found anywhere in the theory of relativity; such a proof must be mathematical and has indeed been given (Reichenbach 1920).

You may still ask what guarantee we have that clocks really behave in a manner consistent with the relativity principle. Well, the most direct proof comes from the study of beams of unstable particles such as mesons. Such a beam constitutes a clock of a kind: the number of particles diminishes along the beam as $\exp(-t/t_0)$ where t_0 is the mean life of the particles and t is their 'proper time', i.e. the time as measured with a clock that moves with them. For instance for (charged) pions $t_0 = 2 \cdot 5 \times 10^{-8}$ sec in which time a pion could travel 25 feet if it went at the speed of light. Without the relativistic slowing-down of clocks, a pion beam would have faded out after a few hundred feet of travel. But with big synchrotrons one can produce pions with $v/c = 0 \cdot 9999$ where the factor γ is about 70; intense pion beams can be obtained extending to many hundred feet from the target, and indeed the decrease in beam intensity along the beam confirms that the pions decay about 70 times more slowly—by laboratory time—than when they are at rest. Less clear-cut but equally striking examples have long been known from cosmic rays (see Janossy 1950).

A somewhat more indirect proof was obtained by Ives and Stillwell (1941). They accelerated hydrogen ions to about 40 kv ($v/c = 0 \cdot 006$); some of the ios then captured an electron and emitted spectral lines, whose Doppler shift was accurately measured. The transverse Doppler effect could not be observed directly because it would have amounted to only $0 \cdot 002$ per cent change in wavelength, which would have been masked by an uncertainty of less than $0 \cdot 2°$ in the direction (relative to the particle stream) of the light observed. Instead they measured the Doppler shifts at two directions forming the same small angle ϵ with the forward and the backward direction of the particle stream; the mean of the two shifts is $\frac{1}{2}(\sqrt{\{(c+v)/(c-v)\}} + \sqrt{\{(c-v)/(c+v)\}})$ $\cos \epsilon = \gamma \cos \epsilon$, and this value was indeed observed.

There is also an experiment by Hay *et al.* (1960) who showed that gamma ray absorption is diminished by fast transversal motion of emitter and absorber; but this experiment employed rotary motion and will consequently be discussed in a later article.

It may also be asked by what mechanism a clock adjusts its rate when its state of motion is altered. That mechanism could probably be worked out for each type of clock, but that would usually be a very difficult exercise, and not really necessary. We have no reason to doubt the accuracy of the relativity principle. Several of its consequences have been accurately tested. Many elastic collisions between fast protons and protons at rest have been seen in cloud and bubble chambers, and the speed and angles of the two protons resulting from such collisions agree with Einstein's formulae; and in proton synchrotrons the magnetic field must be made to increase accurately as a function of the accelerating frequency, a function computed from the relativistic mass increase, or else the synchrotron does not work. In view of all this evidence it seems sensible to use the relativity principle when it offers us a quick answer to a question, just as we use the principle of energy and momentum conservation to predict—in part—the behaviour of a mechanical system without laboriously integrating its equations of motion.

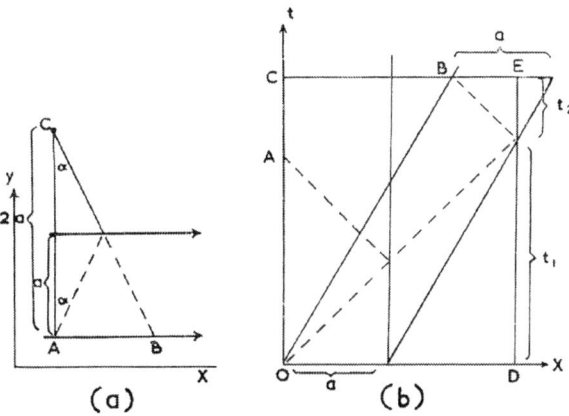

Fig. 7. Time dilatation illustrated by a ' rod clock '. In 7a the bold lines represent the paths of the two ends of a rod of length a, travelling crosswise at the speed v. A light signal, travelling back and forth along the rod, traces out the dotted line, with $\sin \alpha = v/c$. Hence the time interval between the ticks A and B is AC/c . $\cos \alpha = (2a/c)(1-v^2/c^2)^{-\frac{1}{2}} = 2a\gamma/c$, compared to $2a/c$ for the same clock at rest; this is the usual time dilatation.

7b is a space-time diagram, representing the same rod-clock in lengthwise motion, *without* Lorentz contraction, together with one at rest. In the latter, a light signal emitted at O returns at A, after a time $2a/c$. In the moving clock it comes back at B, which is synchronous (in the rest frame) with C. The time interval $OC = t_1 + t_2$; $t_1 = OD/c = (OD-a)/v$, hence $t_1 = a/(c-v)$; $t_2 = BE/c = (a - BE)/v$, hence $t_2 = a/(c+v)$. So $OC = (a/c-v) + (a/c+v) = OA/(1-v^2/c^2) = OA \cdot \gamma^2$. Thus the moving clock seems to be slowed by the factor γ^2 instead of γ. But we must remember that a rod moving lengthwise suffers Lorentz contraction by the factor $1/\gamma$; then the time dilatation comes out correctly.

All the same it may be instructive to consider a very simple type of ' clock ', namely a rod of length a with a flash lamp at each end, so constructed that each lamp will be set off by the flash from the other. Thus the lamps will flash alternately, the interval between two flashes being a/c, in the rod's rest system. If the rod moves at right angles to its own direction, with the speed v, then the light signals—relative to the ' rest ' system—trace out a zigzag line, consequently the intervals between signals will be longer by a factor which can easily be seen (fig. 7a) to be the ' dilatation ' factor γ. But what if the rod moves lengthwise? In that case the simple argument indicated in fig. 7b would seem to show that the clock slows down by the factor $1/\gamma^2$, not $1/\gamma$. But we have forgotten the Lorentz contraction: a rod moving lengthwise at the speed v becomes shorter by the factor $1/\gamma$, speeding up the signals to the correct rate $a/\gamma\text{c}$. This is perhaps the simplest way to show that the relativity principle leads to the Lorentz contraction: if that contraction did not exist, the rate of the ' rod clock ' described would depend on its orientation relative to the direction of motion, in contradiction with the relativity principle.

By the way, the whole last paragraph is nothing but a re-wording of the traditional discussion of the Michelson-Morley experiment: the two arms of the interferometer can be considered as two ' rod clocks ', and the historic experiment established their synchronous behaviour irrespective of orientation.

So we should accept that within the framework of special relativity, i.e. for clocks that for most of the time are at rest in some inertial system, with no large differences in gravitational potential, the terms ' identical clocks ' and ' clocks synchronized by the relativity principle ' are equivalent. That is no longer so with accelerated clocks or in the presence of strong gravitational potential differences, as we shall see in the second article.

REFERENCES

DARWIN, C. G., 1957, *Nature*, **180**, 976.
DINGLE, H., 1956, *Proc. Phys. Soc.* **A69**, 925; 1957, *Nature*, **180**, 1275.
EINSTEIN, A., 1905, *Ann. Physik*, **17**, 891; 1911, **35**, 898.
HAY, J. J., SCHIFFER, J. P., CRANSHAW, T. E., and EGELSTAFF, P. A., 1960, *Phys. Rev. Letters*, **4**, 165.
IVES, H. E., and STILWELL, C. R., 1941, *J. Opt. Soc. Am.*, **31**, 369.
JANOSSY, L., *Cosmic Rays*, p. 22 (Oxford Press 1948 and 1950).
MICHELSON, A. A., and MORLEY, E. W., 1887, *Am. J. Science*, **34**, 333.
MÖSSBAUER, R. L., 1959, *ZS. f. Naturf.*, **14a**, 211.
REICHENBACH, H., 1920, *Relativitatstheorie und Erkenntnis a priori*.
SHERWIN, C. W., 1960, *Phys. Rev.*, **120**, 17.

The Author:

Otto Robert Frisch, O.B.E., F.R.S., is Jacksonian Professor of Natural Philosophy at the University of Cambridge since 1947, and Fellow of Trinity College. Educated in Vienna (Austria); did research in Germany, Denmark, U.K., and U.S.A., mostly on nuclear physics.

Time and Relativity: Part II

by O. R. FRISCH

Jacksonian Professor of Natural Philosophy, Cambridge University

In a previous article we have seen that clocks behave in a somewhat surprising manner, according to the special theory of relativity: two events, simultaneous by the clocks in one inertial frame, will in general not be simultaneous by the clocks in another inertial frame, and of two clocks in relative motion, each will appear slow to an observer moving with the other. But there was no reason to think that two identical clocks at relative rest would not go at the same rate. We shall now show that this is no longer so for clocks in an accelerated system, or in the presence of gravity.

Let us consider a spaceship of length b with its rockets going full blast, giving it an acceleration a; Francis (in front) and Robert (at the rear) have identical clocks. Light takes the finite time $b/\mathbf{c}$ to go from Francis to Robert, and during that time the ship gains the speed $ab/\mathbf{c}$; we shall assume that speed to be small compared with $\mathbf{c}$. By the time Robert receives a signal—say a train of monochromatic light or radio waves—he has the speed $ab/\mathbf{c}$ relative to the inertial system in which Francis was at rest when that signal was sent out. Hence he will observe that signal with a Doppler shift, that is with a frequency increased by the factor $\sqrt{\{(\mathbf{c}+v)/(\mathbf{c}-v)\}} \simeq 1+v/\mathbf{c} = 1+ab/\mathbf{c}^2$. This factor is independent of the frequency of the light and applies equally well to the ticks of Francis' clock; hence that clock will appear fast to Robert, by the factor $1+ab/\mathbf{c}^2$. In the same way we can see that Robert's clock will appear slow to Francis, by the factor $1-ab/\mathbf{c}^2$. All this is a straightforward consequence of the Doppler effect and does not even depend on the principle of relativity.

Inside the ship any object released will no longer be accelerated and will consequently drop behind; it will be seen to move with the apparent acceleration a toward the tail of the ship. Einstein pointed out that this behaviour of objects in the ship might equally well be due to the gravitational attraction of a planet on which the ship was standing on its tail, provided the empirically found proportionality between weight and mass is strictly correct. He took an important step beyond this when he proposed (see Einstein 1911) his Equivalence Principle, by which a uniform gravitational field is equivalent to an accelerated frame of reference in every aspect, including the behaviour of electromagnetic field such as light signals. From this he developed later his 'General Theory of Relativity '; but here we shall be concerned only with the direct consequences of the Equivalence Principle. In particular we can see that two clocks, placed at the top and the foot of a tower, will behave just like our two clocks in the nose and the tail of an accelerating spaceship; we have only to replace b, the length of the ship, by h, the height of the tower, and a, the acceleration of the ship, by g, the acceleration due to terrestrial gravity. The Equivalence Principle then predicts that the observer on the top of the tower would see the other clock go slow compared to his own identical clock by the factor $1-\Delta\phi/\mathbf{c}^2$ where $\Delta\phi = hg$ is the difference in gravitational potential; for instance, excited atoms near the foot of the tower would emit spectral lines which have lower frequency (are shifted toward the red) compared to the lines emitted by the same kind of atoms at the top.

It may be well to make this result plausible in a different way. Let us consider two equal atoms at a different height, connected over a pulley by that useful piece of hardware, a weightless string (fig. 1). We assume that to begin with the lower atom is in its first excited state, possessing the extra energy E and hence—according to the special relativity theory—the extra mass E/c^2. We then allow it to emit that extra energy as a photon which travels upwards until it is absorbed by the upper atom. This will now be the heavier and will sink down, turning the pulley and doing work amounting to $hg \cdot E/c^2 = E \cdot \Delta\phi/c^2$. Now we are apparently again where we started and can repeat the process indefinitely, obtaining work all the time. Have we indeed invented perpetual motion?

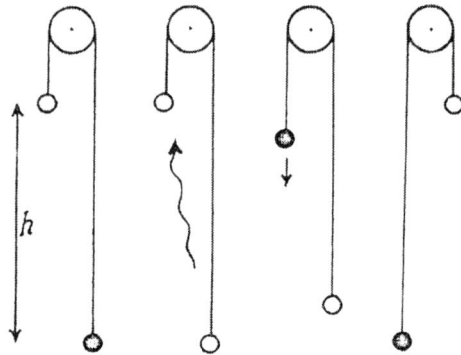

Fig. 1. Two identical atoms, hung over a pulley by a weightless thread, apparently make perpetual motion possible; the puzzle is resolved by the gravitational redshift of photons (see text).

Surely not; the work done by the pulley must come from somewhere. Everything fits if we assume that the photon on arriving at the upper atom has not quite enough energy to excite the upper atom; indeed its energy must fall short just by $E \cdot \Delta\phi/c^2$, the work we expect from the pulley. So the quantum must arrive with the energy $E(1 - \Delta\phi/c^2)$, and if we remember the proportionality between frequency and energy of a light quantum we see that this tallies with the conclusion we got from the equivalence principle.

Actually it is not necessary to bring in the quantum theory; merely from the Lorentz transformation of energy and momentum it can be shown that the energy contained in a light signal changes by the same factor as its frequency if it is observed in a different frame of reference. So the proportionality on which our agreement rests need not be taken from the quantum theory; it is implicit in the special theory of relativity. The fact that the atoms suspended from our weightless string have quantized energy states is not really important; we could have used searchlights instead and would still have to come to the conclusion that a light signal on climbing against gravity loses energy and hence frequency.

But, you will ask, how can the signal lose frequency? Since the distance between source and receiver is constant, surely the number of wave crests that arrive at the receiver must equal the number that left the source during the same time; so the frequency must be the same! That is indeed so provided we use clocks that have been synchronized by signals, say from a common master clock, rather than clocks of identical construction, as we have assumed so far. In the presence of gravity we must distinguish between two ways of establishing time units in places of different gravitational potential. We can (1) set up a system of clocks that are synchronized by wires or radio signals. In that case the frequency of a signal remains constant (by definition!) on travelling from one clock to the other, and any number of clocks at relative rest can be synchronized unambiguously in that way, whatever gravity fields may be present. The only disadvantage of that arrangement is that the frequency of a primary standard such as a caesium clock will no longer be ' standard ': it will depend on the gravitational potential at the point where the clock is placed. Alternatively we can (2) rely on independent clocks of identical construction, for instance caesium clocks; then if we define frequency at each place with reference to the local clock we must accept that a signal decreases or increases in frequency if it travels in a direction of increasing or decreasing gravitational potential.

In practice the differences are so small they don't matter; even on Mount Everest the increase in clock rate against sea level, hg/c^2, would be only one part in 10^{12}. However, caesium clocks and other atomic clocks are now approaching that accuracy, and an experimental test may soon be possible. A much greater difference in potential, $\Delta\phi/c^2 = 2\cdot12 \times 10^{-6}$, exists between the Earth and the surface of the Sun, and even higher potentials must exist on the surface of the very dense white dwarf stars. Einstein predicted as early as 1909 that spectral lines from those sources should be shifted to the red, compared with the lines of the same atoms on Earth. This gravitational red-shift has been looked for most carefully but is obscured by a variety of other line shifts, and although there is some evidence for it the results don't seem to be conclusive.

However, a recent discovery by Mössbauer (1959) has made it possible to use certain radioactive nuclei as extremely accurate ' clocks '. It had long been realized that the frequency of the gamma rays emitted from nuclei is extremely well defined; for instance with a frequency ν around 10^{20} c/sec and a mean life time τ of 10^{-7} sec the natural line width $\Delta\nu/\nu = 1/2\pi\nu\tau$ would be less than 10^{-13}. But the gamma rays actually observed have their frequency reduced by the recoil of the nucleus and the line broadened by its thermal motion (Fig. 2). Mössbauer discovered that when the gamma ray energy is not too high, say below 100 kev, some of the quanta will be emitted with the full excitation energy, the recoil having been taken up by the whole crystal lattice rather than by the emitting nucleus. The lattice is so heavy that its recoil energy is quite negligible, and it can be shown that the thermal broadening also disappears in that case. In absorption, too, the momentum of the quantum can be taken up by the lattice. Both the emission and the absorption spectrum then show a very sharp line superimposed on their normal spectrum (fig. 2), and this gives rise to a strong resonant absorption. Mössbauer showed that this resonance can be destroyed—i.e. the absorber becomes more transparent—if source and absorber are made to move one relative to the other so that the emission line is Doppler-shifted as seen by the absorber; the extreme sharpness of the resonance was strikingly demonstrated by the fact that a relative speed of only 1 cm/sec^{-1} was ample to destroy the absorption!

A search through the known nuclides quickly showed that a particularly favourable nucleus should be ^{57}Fe, whose first excited state—conveniently obtained by the beta decay of ^{57}Co, with a half-life of half a year—lies only 14·4 kev above the ground state and has a long mean life $\tau = 10^{-7}$ sec. The corresponding gamma ray has a frequency $\nu = 3\cdot5 \times 10^{18}$ and a relative line width

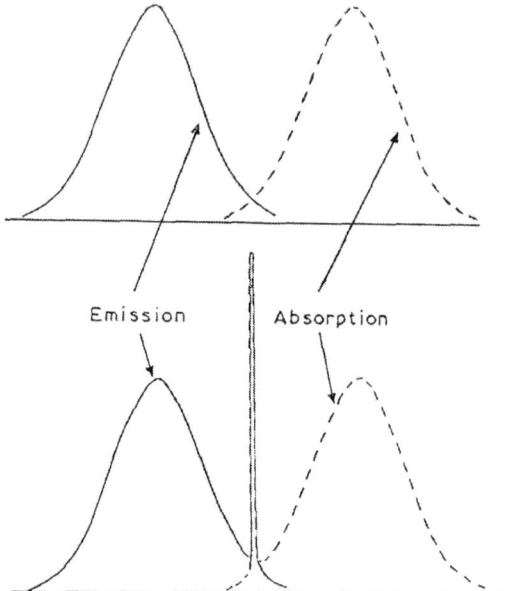

Fig. 2. Both in emitting and in absorbing a gamma quantum of energy E, a free nucleus (mass m) takes up a momentum E/c and hence an energy $E^2/2mc^2$; furthermore the thermal motion causes Doppler broadening of the line. Hence the emission and absorption lines are separated by E^2/mc^2, and both are broadened. But with bound atoms the lattice can occasionally take up the momentum, and the energy lost is then negligible; there appears an unshifted line, both in emission and absorption, causing strong resonance over a width determined chiefly by the mean life of the excited state.

$1/2\pi\nu\tau = 5 \times 10^{-13}$, and from this it follows that the resonant absorption should drop to one half if the source is moved relative to the absorber at a speed of about 0·01 cm/sec^{-1}. Experiments (fig. 3) showed the line to be slightly wider depending on the material used and its heat treatment. The reason is probably, that the line is actually a Zeeman multiplet formed by the effect of the atomic magnetic field on the nucleus, and it is difficult to make the fields exactly alike in source and absorber.

O. R. Frisch

Even so, here was clearly a possibility to detect extremely small frequency changes, and the idea of looking for Einstein's gravitational shift occurred to a number of people. All that seemed necessary was to mount the source and the absorber, one above the other, as far apart as the gamma ray intensity would allow, and then look for a small asymmetry in the Doppler shift curve (fig. 3). A race developed between Harvard and Harwell, in which Harwell obtained the first indication of a shift (Cranshaw et al. 1960); but doubts were cast on that result when it was found by Pound and Rebka (1960) that a shift of the same magnitude might have been produced by a temperature difference of less than 1°c between source and absorber.

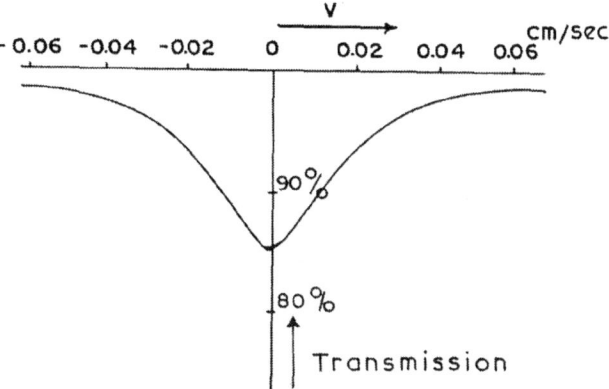

Fig. 3. Typical transmission, through a thin iron foil, of the 14·4 kev gamma rays of excited ^{57}Fe, as function of the relative velocity of source and foil. The absorption is almost entirely due to the ' recoil-less ' resonance, which is so narrow that the Doppler effect from quite slow relative motion is enough to spoil it.

Such a temperature effect had independently been forseen by Josephson (1960) and can be described simply as the transverse (quadratic) Doppler effect due to the thermal motion of the nuclei. The ordinary, linear Doppler effect is of course many thousand times larger, but is absent in the Mössbauer line; one may say that it is cancelled out because the emitting nucleus, carrying out its thermal oscillations, changes its direction many million times during the mean life of the excited state. But there remains the second-order Doppler effect, or in other words the relativistic dilatation, discussed in the previous article. This amounts to $\Delta\nu/\nu = \frac{1}{2}v^2/c^2 = U/2c^2$ if we assume that $\frac{1}{2}v^2$, the mean kinetic energy per unit mass of the lattice, is half its total energy U per unit mass, as it will be if the nucleus is bound to the lattice by elastic forces.

Of course if the absorber is at the same temperature as the source one expects no observable shift, but if they differ by ΔT the shift will be $\Delta T . C_p/2c^2$ where C_p is the specific heat. For iron at room temperature this comes to $2·2 \times 10^{-15}$ for 1°c temperature difference, equivalent to the gravitational shift over a vertical distance of about 70 ft, almost twice the distance used by Cranshaw et al. (1960).

So it is necessary to keep source and absorber at accurately the same temperature. Even then there may be a small shift if source and absorber are not exactly the same material: the Debye temperature and hence the zero point energy may be different, and that may cause a difference in U even if the temperature is the same. By checking all those effects, Pound and Rebka (1960) verified the existence of the gravitational shift beyond doubt and later found it to be 0.97 ± 0.04 times the predicted value.

If the thermal motion caused the line to be red-shifted that means that not all the excitation energy E of the nucleus appears in the radiation emitted; what happens to the missing energy? The answer, surprisingly, lies in the minute decrease in mass, E/c^2, that the nucleus suffers on emitting the quantum. We may consider the nucleus as bound by harmonic forces to its place in the lattice; those forces remain unchanged, and hence its oscillation frequency f—which is proportional to $m^{-1/2}$—will increase by $\Delta f/f = \frac{1}{2}\Delta m/m = \frac{1}{2}E/mc^2$. The fact that the Mössbauer effect is recoilless means that the oscillation quantum number remains unchanged, so the oscillation energy $Q = Um$ of the atom will increase in proportion with the frequency and thus by the amount $\Delta Q = Q \cdot \frac{1}{2}E/mc^2 = E \cdot U/2c^2$, just the amount that we saw was missing. It may be objected that individual atoms do not behave like separate oscillators, but form part of a lattice of many atoms, and that the effect of the change in mass is therefore many times smaller; but then we must also remember that many different oscillations modes of the lattice are affected, and if this is properly allowed for one gets the same answer as before.

It is perhaps instructive to go back to the transverse Doppler effect, discussed in the previous article, and ask what happens to the energy there. Consider an excited atom, with the excitation energy E, travelling at the speed v relative to our ' rest system '. If it emits its energy as a quantum that, in the rest system, travels at right angles to the motion of the atom, that quantum will be red-shifted and will deliver only the energy $E\sqrt{\{1-(v^2/c^2)\}}$ to an absorber at rest. If we ignore powers higher than v^2/c^2 an energy amount $E \cdot \frac{1}{2}v^2/c^2$ is missing. What has happened to the missing energy? Here again the clue is that the atom has become lighter by $\Delta m = E/c^2$. In the rest system its momentum will be unchanged since the photon went off at right angles (ignoring, for the moment, the small recoil effect); so its kinetic energy $E_k = p^2/2m$ has increased by $\Delta m \cdot d(p^2/2m)/dm = E \cdot \frac{1}{2}v^2/c^2$ which is just the missing energy.

Actually with a little algebra we can solve the problem exactly for any speed of the atom, and allowing for the recoil which causes it to change direction. We find that the energy of the emitted photon is changed by the factor $(1 - E/2mc^2)$ $\sqrt{\{1-(v^2/c^2)\}}$. The second term in the first bracket disappears when we neglect the mass E/c^2 of the photon compared to that of the ' clock ', as we usually do in relativity theory. In the same way we can verify that the law of energy conservation holds for the ordinary, linear Doppler effect; but in that case the recoil of the clock, even if it is treated as very heavy, remains significant.

Let us come back to the temperature effect of the Mössbauer line. The measurements (Pound and Rebka 1960) have shown it to be in good agreement with special relativity, and this also shows (Sherwin 1960) that those nuclear clocks are wonderfully insensitive to shaking; their thermal motion subjects them to incessant accelerations of about $10^{16}g$ ($\simeq 10^{19}$ cm/s^{-2}) without affecting their rate by more than one part in 10^{13}.

A more direct test of the effect of speed on clock rate was performed (Hay *et al.* 1960) by having the source near the centre and the absorber at the periphery of a fast spinning wheel. Speeds up to about 200 m/s were used, giving an expected time dilatation by about 2 parts in 10^{13}, nearly one half of the line width for ^{57}Fe. The difficulty here was to avoid the linear Doppler effect, which in the line of motion would have been about a million times larger. It would be impossible to mount the source so close to the centre of the wheel that the absorber would have no significant radial velocity component relative to the source. The problem was solved by attaching the source to the wheel, near its centre; then as long as the wheel is rigid enough (and this was checked) the distance between any parts of the absorber and of the source remains constant despite the rotation, and thus there is no first-order Doppler effect. Figure 3 shows that for small speeds the increase in transmission goes with the square of the line shift, which in turn goes with the square of the speed. The resulting increase of transmission with the fourth power of the speed was indeed observed, up to the calculated increase by about 4 per cent at 200 m/s. Of course the direction of the line shift could not be verified in this experiment, only the amount.

It is quite legitimate to use special relativity as long as we refer all motions to an inertial system, most conveniently to that in which the centre of the wheel is at rest. But what if we use a frame of reference that rotates with the wheel? In that frame both source and absorber are at rest; but an observer experiences a centrifugal acceleration $r\omega^2$ which he attributes to a gravitational force with the potential $\frac{1}{2}r^2\omega^2 = \frac{1}{2}v^2$. Consequently if the observer is at the edge of the wheel he expects photons, emitted near the centre, to be blue-shifted by $\Delta\nu/\nu = \phi/c^2 = \frac{1}{2}v^2/c^2$ on arriving at the periphery. But this is just what we deduced from special relativity, using an inertial frame of reference. The experiment of Hay *et al.* has sometimes been said to confirm a conclusion from general relativity; but actually special relativity is quite sufficient to predict its outcome, and the equivalence principle need not be invoked.

This is perhaps a good place to go back to the twin paradox, as discussed in the previous article. There we concluded that Jack, who has been at rest in the same inertial frame all the time, experienced the passage of a longer time than his twin brother Albert who travelled out into space and back again. But are we not entitled to use Albert's ship as our frame of reference? In that frame, Albert is stationary throughout while it is Jack who moves out into space and back again. So now it looks as if it should be Albert, not Jack, whose clock shows the passage of a longer time. Since those two answers are in conflict it has sometimes been argued that neither can be true, and that both brothers must age the same, whichever way they move.

But if we choose a frame in which Albert is at rest and Jack goes out and back again, then in that frame there must be a gravitational field to account for the accelerations experienced by Albert, and for the fact that Jack does not feel any acceleration although he goes out and back. We shall show that if we include the frequency shifts produced by those gravitational fields we arrive at the same conclusion as before, and that there is no conflict between the two treatments.

To simplify the argument we shall assume again that the speeds are so small that powers higher than $(v/c)^2$ can be neglected. The initial speed-up and the final slow-down are unimportant because Jack and Albert are so close together that they are both at almost the same gravitational potential (and might indeed avoid those accelerations by comparing their clocks in flight as they pass each other without matching speed, neither initially nor at the end). The important period is that during which Jack—in Albert's system!—reverses speed. Let us say that Albert puts on his rockets for the time t so as to produce an acceleration a; then in his frame of reference there is a corresponding gravitational field that counteracts the rockets and, during the time t, gives the acceleration a to Jack. The distance between the brothers at that period is vT where T is the time they have been flying apart at the speed v. (In our approximation it does not matter whether T is measured by Jack or Albert.) Therefore the gravitational potential difference between them is avT, and the gravitational shift causes Jack's clock to gain the amount $tav\ T/c^2$ during the time t the acceleration is on. Now t has to be long enough to cause reversal, that is $a\,t\,=\,2v$. Hence the time gained by Jack's clock is $2T\ v^2/c^2$. But this is just twice the amount which it should have lost according to the argument from the previous article; remember that in Albert's frame, which we have now been using, it is Jack who has been the traveller. Altogether Jack's clock has gained Tv^2/c^2 and this is just the amount we found previously by referring everything to Jack's inertial system. So the answer is the same whichever of the two frames of reference we use.

Of course we have demonstrated this only for small values of v/c where powers higher than $(v/c)^2$ can be neglected. Otherwise we would have needed the mathematical apparatus of general relativity; that demonstration has been published by Born and Biem (1958).

It remains to say a little about satellites. Circling the Earth once in 90 min, a low-orbit satellite has a speed of $7\cdot7 \times 10^5$ cm/s and hence the time dilatation amounts to $3\cdot3$ parts in 10^{10}, an amount easily measurable with present-day atomic clocks (Singer 1956). On the other hand, the gravitational field between us and the altitude at which the satellite travels speeds up its clock rate. If we confine ourselves to circular orbits of radius r while r_0 is the radius of the Earth, both the velocity v and the gravitational potential ϕ are constant over the orbit; the relative speeding-up of the satellite clock compared to an identical clock stationary at sea level is then $\Delta\nu/\nu=(\Delta\phi - v^2/2)/c^2$. Now $\Delta\phi = \int_{r_0}^r (gr^2_0/r^2)dr$ $=gr^2_0(1/r_0-1/r)$. To get v we write the centripetal acceleration $v^2/r =gr^2_0/r^2$, hence $v^2=gr^2_0/r$. So we get $\Delta\nu/\nu=(gr_0/c^2)\,(1-3r_0/2r)=7 \times 10^{-10} \times (1-3r_0/2r)$. For low altitudes the time dilatation (the second term) prevails; above an altitude of $r_0/2$ (just about 2000 miles) $\Delta\nu/\nu$ becomes positive because the first term, the relativistic blue-shift, becomes dominant. These predictions will presumably be tested within the next few years though there would seem to be little doubt about the outcome.

REFERENCES

BORN, M., and BIEM, W., 1958, *Kon. Ned. Akad. Wet.* **B61**, 110.
CRANSHAW, T. E., SCHIFFER, J. P. and WHITEHEAD, A. B., 1960, *Phys. Rev. Letters*, **4**, 163.
EINSTEIN, A., 1911, *Ann. d. Physik*, **35**, 898.
HAY, H. J., SCHIFFER, J. P., CRANSHAW, T. E., and EGELSTAFF, P. A., 1960, *Phys. Rev. Letters*, **4**, 165.
JOSEPHSON, B. D., 1960, *Phys. Rev. Letters*, **4**, 341.
MOSSBAUER, R., 1959, *Z. Naturforsch.*, **14a**, 211.
POUND, R. V. and REBKA, G. A. Jr., 1960, *Phys. Rev. Letters*, **4**, 274, 337;
SHERWIN, C. W., 1960, *Phys. Rev.*, **120**, 12.
SINGER, S. F., 1956, *Phys. Rev.*, **104**, 11.

Date Due

CPSIA information can be obtained
at www.ICGtesting.com
Printed in the USA
LVHW081605191022
731075LV00004B/187

9 781013 449833